Gordon A. Parker

Analytical Chemistry of
Molybdenum

With 45 Tables

Springer-Verlag
Berlin Heidelberg New York Tokyo 1983

Gordon A. Parker
Departement of Chemistry
University of Toledo
Toledo, OH 43606/USA

ISBN-13: 978-3-642-68994-9   e-ISBN-13: 978-3-642-68992-5
DOI: 10.1007/978-3-642-68992-5

Library of Congress Cataloging in Publication Data.
Parker, Gordon A., 1936–. Analytical chemistry of molybdenum. Bibliography: p. Includes index.
1. Molybdenum – Analysis. I. Title.
QD181.M7P3 1983   546'.5346   83-504

2152/3020-543210

Contents

Chapter 1

Introduction

The importance of molybdenum, in industry, in life processes, etc., necessitates accurate and precise analytical methods to determine its presence and its concentration in a variety of samples. With different samples molybdenum is present, as a major constituent, as a minor constituent, as one of several alloying components, as a trace element, and possibly in different oxidation states. The analyst is required to develop more selective and sensitive procedures for determining molybdenum in any sample type and new and improved methods for molybdenum are appearing in the literature at an ever increasing rate.

It is the purpose of this work to summarize presently available methods for determining molybdenum. It proceeds in two parts. The first discusses procedures by technique, precipitation, colorimetry, electrochemistry, etc. The second discusses molybdenum in specific materials, alloys, biological, environmental, etc. In this latter portion emphasis is placed upon sample treatment to obtain a working solution which can then be processed by the most expedient of the techniques discussed earlier.

Throughout this work explanation is presented to provide background for a particular analytical approach. Selected laboratory procedures are then given in detail enabling one to carry out the experiment. Extensive tables accompany most chapters referring one to additional sources of analytical methods applied to specific molybdenum-containing samples. Selection of procedures and additional references is by no means complete but rather based upon availability, applicability, clarity, and acceptability of results. Every attempt has been made to eliminate errors in preparing the experimental instructions, however, should some be found please call them to the author's attention.

Certain techniques applied to determination of molybdenum are omitted because of their limited application. Other techniques, because they require highly specialized skills and expensive instrumentation, are mentioned only briefly.

It is not sufficient to follow, blindly, the procedures given here. Every sample requires individual study and a knowledge of possible interfering constituents. The reader is obliged to draw from several chapters to achieve proper dissolving procedure, to isolate molybdenum from its interferences or in some way sequester these interferences without separation, to select the appropriate analytical technique based upon available equipment, operator experience, and sensitivity; and, finally, to properly calculate and report the results of his/her findings. Considerable chemical knowledge is needed to satisfactorily weight these points and to respond to unexpected observations based upon error or overlooked details. Experimental skill is also necessary if a proper determination of molybdenum is to be achieved.

Several earlier comprehensive reviews of the analytical chemistry of molybdenum are available and should be consulted for insight into the analysis of molybdenum-containing materials [1–6]. In addition the current journal literature provides additional pertinent information.

References

1. Busev, A.I.: Analiticheskaya Khimiya Molibdena 1962 Moscow, Izd. Akad. Nauk SSSR
2. Elwood, W.T., Wood, D.F.: Analytical Chemistry of Molybdenum and Tungsten, 1971, Oxford, Pergamon Press
3. Fresenius, R., Jander, G.: Handbuch der analytischen Chemie Part II Vol. 6, 1948 Berlin, Springer-Verlag
4. Gmelins Handbuch der anorganischen Chemie, Vol. 53, 8 th edition 1935, Berlin, Verlag Chemie
5. Mann, J.W.: Encyclopaedia of Industrial, Chemical Analysis Vol. 16 (Snell, F.D., Hilton, C.L. ed.) 1972, New York, Wiley-Interscience
6. Parker, G.A.: Treatise on Analytical, Chemistry Part II Vol. 10 (Kolthoff, I.M., Elving, P.J. ed.) 1978 New York, Wiley-Interscience

Chapter 2

Detection

Qualitative identification of a component in an unknown matrix has traditionally meant chemical testing. The classical hydrogen sulfide scheme frequently gives way, however, to spot testing and to instrumental means of identification through emission spectroscopy or x-ray fluorescence measurements. These latter instrumental techniques are faster, can be carried out in conjunction with quantitative measurements, and are frequently automated providing, in some cases, almost instantaneous results. If suitable instrumentation is lacking, chemical tests are employed. Details regarding spectroscopic and x-ray fluorescent procedures are discussed in subsequent chapters. Chemical methods of testing for molybdenum are presented here, briefly, and in greater detail elsewhere [1–3].

1. Detection in the Presence of Many Ions

Those choosing to carry out the traditional qualitative procedure will find molybdenum in the acid insoluble sulfide group along with Cu, Pb, Cd, Hg, Bi, As, Sb, Sn, and others [4–7]. After precipitation with hydrogen sulfide in the presence of dilute acid, 0.3 M hydrochloric acid, the solution is partially neutralized with ammonia. Mo, As, Sb, and Sn dissolve as polysulfides and are separated from the copper group metals. Addition of 5 M HCl and boiling to expel H_2S results in As and Mo precipitation as sulfides while Sb and Sn remain in solution. Following filtration, molybdenum and arsenic are dissolved in dilute nitric acid, the solution neutralized with ammonia and magnesium carbonate added. Arsenic precipitates under these conditions, molybdenum does not. Molybdenum in the filtrate is determined by a suitable confirmatory test.

Paper chromatography separation of numerous cations followed by application of suitable spray reagents is a convenient means for identifying molybdenum when present with several other cations. The procedure of Schneer-Erdey separates 42 cations on a 30 cm strip of chromatography grade paper (Schleicher and Schüll, number 2043/A) [8]. Details of this procedure are:

Place a few drops of sample solution dissolved in dilute hydrochloric acid, HCl, near the bottom edge of a 2×30 cm strip of S & S No. 2043/A chromatography paper. After the spot has dried, suspend the strip in a suitable covered cylinder which has been lined with filter paper and saturated with a developing solvent consisting of ethanol-12 M HCl-water (15: 4: 1 v/v). The spot should be just above the surface of the developing solvent present in the bottom of the cylinder. Cover the cylinder and allow the chromatogram to develop for several hours until the solvent front is near the top of the paper strip. Remove the strip from the cylinder, mark the location of the solvent front, and allow the paper to air dry. Under these conditions molybdenum has an R_f value of 0.7, that is the molybdenum spot will be located 0.7 of the distance traveled by the solvent front. In the region of $R_f = 0.7$, spray the strip with an 0.2% (w/v) solution of morin, $2',3,4',5,7'$-pentahydroxyflavone $C_{15}H_{10}O_7$, in 95% ethanol. A brown color which, when viewed with an ultraviolet lamp, appears even more intensely brown indicates the presence of molybdenum. Detection limit for this procedure is 0.15 µg molybdenum.

The procedure of Poonia [9] uses a circular piece of chromatography paper rather than a paper strip to separate 22 metal ions. An 11 cm disk of chromatography paper (Whatman number 1) is cut to form a wick which when bent projects downward from the center of the paper disk into a developing solvent. The disk is supported upon the rim of the beaker or other suitable container in which the developing solvent is placed. The sample is spotted in the center of the paper and development is radial from the center of the paper outwards in all directions. Details of the procedure are:

Place a few drops of the sample in water or dilute acid upon the center of an 11 cm circular Whatman number 1 paper disk which has been afixed with an appropriate paper wick. After the spot has dried, support the paper on the rim of a beaker filled to the level at which the wick dips into the developing solvent. Invert a larger beaker over the assembly to help reduce solvent evaporation from the paper surface. The developing solvent is chloroform-acetone-isopentyl alcohol-11 M hydrochloric acid (1:1:1:0.5 v/v). Development time is about 4 h. After the solvent has nearly reached the edge of the paper, remove the paper, discard the wick, and mark the advance edge of the solvent front. When the paper is dry, spray it with an 0.1% (w/v) solution of potassium ferrocyanide, $K_4Fe(CN)_6$. R_f value for molybdenum under this condition is about 0.8, that is, the molybdenum spot will be located 0.8 of the distance traveled by the solvent front. A reddish-brown spot at R_f approximately 0.8 indicates the presence of molybdenum. U(VI) with R_f slightly less than 0.7 and copper(II) with R_f approximately 0.45 also yield brown colored spots that might be confused with molybdenum. Detection limit for molybdenum by this procedure is not listed.

2. Spot Tests

There is no one reagent that when added to a solution containing molybdenum produces a unique response irregardless of other ions present. If one has preliminary information regarding the composition of a test solution, separation prior to spot testing for molybdenum is desirable. Spot tests are widely used and satisfactory if one realizes that the observed precipitate and/or color could be due to ions other than molybdenum. Some knowledge of the reaction products of the test reagent both with molybdenum and other ions commonly associated with molybdenum is necessary to properly interpret the observed results.

Most common spot test for identifying molybdenum is formation of the orange-red molybdenum(V) thiocyanate complex in the presence of strong acid solution [10–12]. In the presence of 2 M H^+ or stronger acid solution, tin(II) is added to the test sample followed by addition of thiocyanate ion. An orange-red color is observed if molybdenum is present. Iron under these conditions is reduced to Fe(II) which does not produce a red color with thiocyanate. Metals that precipitate with thiocyanate ion will cause turbidity which may prevent one from observing the molybdenum thiocyanate color.

Bermejo-Martinez and Souza-Castelo [13] prefer tin(II) as the reducing agent for molybdenum when it is dissolved in glycerol-ethanol (3:1 v/v) or various glycol solvents. Stability of Sn(II) in these solvents is considerably improved over what one finds in concentrated hydrochloric acid. Their procedure of qualitative molybdenum identification is:

Table 2.1. Qualitative Detection of Molybdenum by the Addition of Various Reagents

Reagent	Condition	Observation	Sensitivity µg	Interference	Ref.
α-benzoinoxime, 2% (w/v) EtOH	4 M HCl	white precipitate	–	Cr, Fe, V, W	[15]
2,2'-bipyridine, 3% (w/v) EtOH	H^+, Sn(II)	red-violet	10	W	[16]
ethylxanthate K^+, 1% (w/v) aqueous	0.016 M HCl	red-violet	0.04	Fe	[17, 18]
gossypol, 0.1% (w/v) EtOH	H^+	red	0.1	Sn	[19]
o-hydroxyphenyl-fluorone, 0.1% (w/v) EtOH	treated paper, spot with sample	red	30	Cr, Cu, Mn	[20]
2-mercapto-3(2-furyl)-propenoic acid, 1% (w/v) EtOH	0.1 M H^+	red	–	Cr, Mn, Ni, V	[21, 22]
α-nitroso-β-naphthol, 1% (w/v) EtOH	H^+	red precipitate	–	V	[23]
phenylhydrazine, saturated 1 M H_2SO_4	H^+	red precipitate	25	Pb, Sn	[24, 25]
tetramethyl-diaminodiphenyl-methane, 0.5% (w/v) 0.5 M HOAc	pH 7	blue precipitate	–	V, W	[26]
thiocyanate K^+, 20% (w/v) aqueous	2 M H^+	orange-red	0.05	Cu, Hg	[13]
toluene-3,4-dithiol, 0.5% (w/v) 1 M NaOH	2 M HCl	green precipitate	–	Re, W	[27]
xylenol orange, 0.1% aqueous	NH_2OH 0.01 M HCl	red-orange	0.02	EDTA	[28]

Place one drop of the sample solution in a small test tube. Add one drop of 6 M hydrochloric acid, HCl. Add 10 drops of 20% (w/v) potassium thiocyanate, KSCN. Add 5 drops 5% (w/v) tin(II) chloride in glycerol-ethanol 3:1 (v/v). (Prepare by dissolving 5 g of $SnCl_2$ in 75 mL of a mixture of 3 parts glycerol, $C_3H_8O_3$, plus one part 95% ethanol, C_2H_5OH. Avoid excessive stirring which traps air bubbles in the solution. When dissolved, add additional solvent mixture until the final volume is 100 ml.) Mix the contents in the test tube. Appearance of an orange-red color indicates molybdenum. Detection limit for this procedure is 0.05 µg Mo.

A similar test forms the molybdenum(V)-thiocyanate-1,10-phenanthroline complex which when extracted into benzyl alcohol produces a pink color [14]. Tungsten, vanadium, cobalt, and copper interfere. Various other spot tests for molybdenum are listed in Table 2-1.

3. Detection in Specific Materials

Providing one accounts for possible interfering ions also present in a sample, any of the tests described is suitable for identifying molybdenum. There are, however, certain procedures specifically adopted to molybdenum in specified environments, namely ores and steels. A simple and quick test for molybdenum in ores is to grind a powdered sample of the ore with sodium thiosulfate, $Na_2S_2O_3$. Potassium hydrogen sulfate, $KHSO_4$, is then added and the sample is again ground. A red-brown color different from that of the original ore is observed if molybdenum is present [29, 30]. Toluene-3,4-dithiol has also been applied specifically to the detection of molybdenum in ores [31]. Filter paper placed upon a metal surface previously moistened with 8 M nitric acid, HNO_3, will, if molybdenum is present, reduce the molybdenum to a lower oxidation state. The molybdenum blue color of these lower oxides is observed on the paper [32, 33]. Tungsten and other easily reduced metals also form reduced oxides and interfere. Potassium ethylxanthate [34–36] and gossypol [37] have been used specifically for detecting molybdenum in steels.

References

1. Fresenius, R., Jander, G.: Handbuch der analytischen Chemie, Part II, Vol. 6, 1948, Berlin, Springer-Verlag
2. Gmelins Handbuch der anorganischen Chemie, Vol. 53, 8th edition 1935, Berlin, Verlag Chemie
3. Feigl, F.: Spot Tests in Inorganic Analysis, 6th Edition, 1972, Amsterdam, Elsevier
4. Alstodt, B.S., Benedetti-Pichler, A.A.: Ind. Eng. Chem., Anal. Ed., 11, 294 (1939)
5. Miller, C.C.: J. Chem. Soc. (London), 786 (1941)
6. Porter, L.E.: Ind. Eng. Chem., Anal. Ed., 6, 138 (1934)
7. Holness, H., Lawrence, K.R.: Analyst (London), 78, 356 (1953)
8. Schneer-Erdey, A.: Talanta, 10, 591 (1963)
9. Poonia, N.S.: J. Chem. Educ., 44, 477 (1967)
10. Kedesdy, E.: Mitt, Kgl. Materialprüfungsamt, 31, 173 (1913); Chem. Abstr., 7, 3940[1] (1913)
11. Krauskopf, F.C., Swartz, C.E.: J. Am. Chem. Soc., 48, 3021 (1926)
12. Rao, D.V.R.: Curr. Sci., 21, 257 (1952)
13. Bermejo-Martinez, F., Souza-Castelo, M.D.P.: Microchem. J., 16, 94 (1971)
14. Rao, V.P.R., Rao, K.V., Sarma, P.V.R.B.: Mikrochim. Acta, (1), 89 (1968)
15. Charlot, G., Bezier, D., Gauguin, R.: Rapid Detection of Cations, 1954, New York, Chemical Publishing Comp.
16. Komarovskii, A.S., Poluektov, N.S.: J. Appl. Chem. USSR, 10, 565 (1937)
17. Koppel, J.: Chem.-Ztg., 43, 777 (1919)
18. Pavelka, F., Laghi, A.: Mikrochem. Ver. Mikrochim. Acta 31, 138 (1943)
19. Vioque-Pizarro, A.: Anal. Chim. Acta, 6, 105 (1951)
20. Gillis, J., Claeys, A., Hoste, J.: Anal. Chim. Acta, 1, 421 (1947)
21. Izquierdo, A., Calmet, J.: Quim. Anal., 28, 148 (1974)
22. Izquierdo, A., Calmet, J.: Analusis, 4, 200 (1976)
23. Shemyakin, F.M., Belokon, A.N.: C. R. Acad. Sci. URSS, 18, 277 (1938); Chem. Abstr., 32, 4467[5] (1938)
24. Montignie, E.: Bull. Soc. Chim. Fr., [4], 47, 128 (1930)

25. Rovira, L.: Rev. Fac. Cienc. Quim. Univ. Nac. La Plata, *16*, 235 (1941); Chem. Abstr., *36*, 6108^6 (1942)
26. Papafil, M., Cernatesco, R.: Ann. Sci. Univ. Jassy, *15*, 384 (1929); Chem. Abstr. *23*, 3187^9 (1929)
27. Clark, R.E.D., Neville, R.G.: J. Chem. Educ. *36*, 390 (1959)
28. Hsu, P.Y., Wang, C.Y.: Hua Hsueh Hsueh Pao, *31*, 264 (1965); Chem. Abstr., *63*, 15531c (1965)
29. Isakov, P.M.: Nauch. Byull. Leningrad. Gos. Univ., A. A. Zhdanova, (33), 31 (1955); Chem. Abstr., *54*, 11844c (1960)
30. Lipchinski, A., Kr'steva, M.: God. Khim.-Tekhnol. Inst., *4*, 13 (1957); Chem. Abstr., *55*, 14170f (1961)
31. Stubbs, M.F.: Analyst (London), *93*, 59 (1968)
32. Getzov, B.B.: Zavod. Lab., *4*, 583 (1935)
33. Neace, J.C.: Chemist-Analyst, *50*, 77 (1961)
34. Leïba, S.P., Shapiro, M.M.: Zavod. Lab., *3*, 503 (1934)
35. Evans, B.S., Higgs, D.G.: Analyst (London), *70*, 75 (1945)
36. Reboul, P.: Chim. Ind., *64*, 574 (1950)
37. Vioque-Pizarro, A., Malissa, H.: Mikrochem. Ver. Mikrochim. Acta, *40*, 396 (1953)

Chapter 3

Separations

Certain reagents combine only with molybdenum when selected other ions are present but no reagent is specific only for molybdenum in the presence of all possible interfering ions. Either one selects a procedure uneffected by other ions known to be present in a sample or in some way isolates molybdenum so the effect of other ions is eliminated or at least minimized. Frequently isolation of a desired constituent from a complex matrix takes the form of a separation. Precipitation, extraction, ion exchange, and other separation techniques are well known and necessary if interfering ions are present. Removal of the desired constituent is beneficial too if only a trace amount of the material is present in a large sample volume. By condensing the element of interest into a smaller volume, thus increasing its concentration, techniques which might otherwise be insensitive can be employed.

Specific separations prior to determination of molybdenum are mentioned throughout this book in connection with various techniques and applications cited in the following chapters. A general overview of molybdenum separation is given in the present chapter.

1. Precipitation

The well known sulfide scheme of qualitative identification frequently serves as a starting point for separations prior to quantitative determination. Molybdenum, belonging to the acid insoluble sulfide group, of this scheme is precipitated, after removal of the insoluble chloride group, from 0.3 M hydrochloric acid solution by addition of hydrogen sulfide. Separation from other acid insoluble sulfides is discussed in the previous chapter, chapter 2, in connection with the qualitative identification of molybdenum. For quantitative separation both double precipitation, to remove all traces of molybdenum, and through washing of the precipitate, to remove adsorbed impurities, are necessary. Molybdenum, in the form of molybate ion, is soluble in basic solution. Precipitation of insoluble hydrous molybdenum(VI) oxide is possible but, because of molybdenyl ion formation in acid solution, requires careful pH control. Foreign ions, too, tend to adsorb upon the freshly precipitated hydrous molybdenum oxide. Various precipitating agents for molybdenum, both inorganic and organic, are discussed in chapter 4.

Trace amounts of molybdenum are sometimes collected from large sample volumes by co-precipitation using an appropriate insoluble carrier which adsorbs molybdenum ions. The use of carriers is commonplace in many radiochemical procedures and finds increasing use in concentrating trace elements from natural waters. Various insoluble hydrous oxides [1] and other insoluble substances [2, 3] co-precipitate molybdenum. There is an optinum pH at which the maximum amount of molybdenum is adsorbed on a precipitating substance. With hydrous iron(III) oxide carrier the optimum pH is from 4–5 [4]. With manganese dioxide a pH of 2.0 is best [5]. Hydrous zirconium oxide also precipitates molybdenum if the pH is not too high [6, 7]. Molybdenum, specifically, is separated from chromium (by co-precipitation of Cr on $Mg(OH)_2$ at pH 12) [8], Fe ($Fe_2O_3 \cdot xH_2O$ precipitates, Mo not adsorbed if NaCl is present and pH > 9) [9], Re (by co-precipitation of Mo on $Fe_2O_3 \cdot xH_2O$ at pH < 7.5) [10], Re (by precipitation of Mo(V) hydrous oxide at pH 5.5) [11], V (by precipitation of MoS_3) [12], and W (by precipitation of MoS_3 in the presence of formic acid) [13].

Molybdenum is separated from rhenium by internal electrolysis under either acid or base condition. Molybdenum is deposited at the cathode while rhenium remains in solution [14–16]. With a mercury cathode, molybdenum is separated from iron and vanadium in acetate buffer, pH 5–6, at a temperature of about 80 °C [17].

2. Extraction

Molybdenum(VI), in strong acid solution, exists as an ion complex rather than the simple molybdenyl ion.

$$MoO_2^{2+} + nL^{x-} \rightarrow MoO_2L_n^{2-nx}.$$

With hydrochloric acid solutions, L represents Cl^- ion. These complexes are extractable by oxygen containing organic solvents, ethers, esters, alcohols, and ketones. Numerous studies pertaining to optinium acid strength and choice of extracting solvent have been made. Generally, interest has centered upon removal of molybdenum from pure solutions rather than from real samples where interfering ions could also be extracted [18–22].

The molybdenum(V) thiocyanate complex, often used for colorimetric measurement of molybdenum, is also extracted by oxygen containing organic solvents. Diethyl ether and methyl isobutyl ketone are commonly employed for this purpose. Ternary complexes, containing thiocyanate and an additional ligand, are soluble in chloroform and non-oxygen containing solvents. Table 6-1 in the chapter on colorimetry lists conditions for extracting molybdenum as a ternary molybdenum thiocyanate auxiliary ligand complex.

Heteropoly molybdenum species, notably molybdophosphate heteropoly ions [23] and reduced molybdophosphate ions [24, 25] are extracted from acid solutions with i-butanol, methyl isobutyl ketone, and other solvents. Addition complexes between molybdophosphate ion and di-n-octylamine [26] or quinoline [27] allow the former to be extracted with chloroform and the latter with nitrobenzene.

Tributyl phosphate [28–31], other substituted phosphates [32–34] and arsenic containing reagents [35] extract molybdenum from acid solutions. Amine containing ligands also combine with molybdenum to form extractable complexes. Aniline [36], Aliquat 336 [37], and the liquid anion exchange resin Amberlite LA-2 [38] are all satisfactory for this purpose. 8-Hydroxyquinoline [39] and substituted 8-hy-

Table 3.1. Extraction of Molybdenum Prior to Colorimetric Measurement of the Extracted Species

Reagent	Condition	Separation From	Ref.
alizarine red S, 0.0025 M dichloroethane	pH 5, acetate, 0.025 M zephiramine	Co, Cu, Th	[49]
benzoylacetone, 0.15 M CHCl$_3$-n-BuOH 1:1	0.01 M HCl, extract 15 min	Cr, Fe, Ni	[50]
N-benzoylphenylhydroxylamine, 0.1 M aqueous	pH 2–3, HCl, extract hexanol	–	[51]
1,4-dihydroxyphthalimide dithiosemicarbazone 0.03% (w/v) DMF	pH 2.6, ClCH$_2$COOH-NaOH, extract i-pentanol	W	[52]
ethyl xanthate K$^+$, 1% (w/v) aqueous	3 M HCl, extract acetophenone	Fe	[53]
3-hydroxy-2-methyl-1-phenyl-4-pyridone, 0.0002 M CHCl$_3$	1 M HCl	Fe, Ga, Ti	[54]
2-mercapto-3-(2-furyl)propenoic acid, 1% (w/v) EtOH	0.1 M HCl, extract CHCl$_3$	Cr, Cu, Ni, V	[55]
o-nitrophenylfluorone, 0.003 M EtOH+dianti-pyrinylmethane, 0.1 M CHCl$_3$	1 M HCl, extract CHCl$_3$	Cr, Fe, V, Zn	[56]
pyrocatechol, 0.3 M aqueous+ triphenyltetrazolium Cl$^-$, 0.4 M aqueous	pH 0.5, HCl+KCl, extract C$_2$H$_2$Cl$_2$	Cd, Co, Fe, Ni	[57]
pyrocatechol-3,5-disulfonate Na$^+$, 0.1 M aqueous+ diphenylguanidine Cl$^-$, 1 M aqueous	pH 2.2, HCl+KCl, extract pentanol-CHCl$_3$ 1:1	Fe	[58]
l-pyrrolidinecarbodithioate NH$_4$$^+$, 0.1 M aqueous	pH 4.8, acetate, extract CHCl$_3$	–	[59]
thiolactic acid p-phenetidide, 2% (w/v) EtOH	0.01 M HCl, extract i-pentanol-benzene 1:1	W	[60]

droxyquinolines [40] extract molybdenum from acid solution with several organic solvents [41]. Other reagents used to extract molybdenum include α-benzoinoxime [42], N-benzoylphenylhydroxylamine [43], diethyldithiocarbamate [44, 45], dithizone [46], ethyl xanthate [47], and ammonium l-pyrrolidinecarbodithioate [48]. Reagents for extraction of molybdenum prior to atomic absorption measurement are listed in Table 8-1 of chapter 8 and for separation of ^{99}Mo from fission products in chapter 12. Table 3-1 of this chapter lists specific reagents which, in addition to separating molybdenum from interfering ions, serve for colorimetric measurements of the extracted species. Table 3-2 lists other extraction procedures for molybdenum.

Following is the procedure of Yatirajam and Ram for separating molybdenum as the reduced molybdophosphate ion [25]. From dilute sulfuric acid solution and

Table 3.2. Separation of Molybdenum from Selected Ions by Extraction

Reagent	Condition	Separation From	Ref.
N-benzoylphenyl-hydroxylamine, 0.02 M EtOH-H_2O, 3:1	pH 3, 0.1 M tartaric acid, extract $CHCl_3$	W	[61]
di-n-butylphosphate, 0.5 M gasoline	0.5 M HCl+0.1 M EDTA+0.1 M ascorbic acid, back extract 0.1 M HCl+1.5% (v/v) H_2O_2	Ga, In, Ti	[62]
diethyldithiocarbamate Na^+, 0.1% (w/v) aqueous	2 M H_2SO_4+2% (w/v) citric acid, extract $CHCl_3$	Nb, Ti, V, W	[63]
bis(2-ethylhexyl)phosphate, 0.1 M kerosine	0.1 M HNO_3+0.1 M NH_4NO_3	W	[64]
mono(2-ethylhexyl)phosphate, 0.1 M C_6H_6	1 M HCl, back extract 0.05 M H_2O_2	Cr, Co, Cu, Ni	[65]
ethyl xanthate K^+, 0.4 M aqueous	2 M HCl+N_2N_4, boil, 10 M HCl, extract $CHCl_3$	U	[66]
hexyldiantipyrylmethane, 0.125 M $C_2H_2Cl_2$	8 M HCl+2 M HF	W	[67]
hydroxyphenylmethyl-phosphonic acid diisopropyl ester, 6% (w/v) CCl_4	6 M HCl	Re	[68]
–	2 M HCl, citric acid, extract acetylacetone-$CHCl_3$, 1:1	W	[69]
–	$N_2H_4 \cdot 2HCl$, boil, 6 M HCl+ Br_2 dropwise, extract pentyl acetate	Re	[70]
–	$N_2H_4 \cdot 2HCl$, boil, 7 M HCl, extract pentyl acetate	V	[71]
H_3PO_4, 0.03 M	$N_2H_4 \cdot H_2SO_4$+0.25 M H_2SO_4, boil, extract MIBK	Fe, U, V, W	[25]
tributylphosphate, 0.5 M C_6H_6	0.1 M H_2SO_4+1.5% (v/v) H_2O_2	Cr, Nb, Ti, V	[72]
tributylphosphate, 0.6 M CCl_4	4 M HCl	Te	[73]
trioctylamine, 0.05 M toluene	0.05 M H_2SO_4	Cr, Fe, V	[74]
trioctylamine, 4% (v/v)+ Aliquat 336, 0.1% (v/v) kerosine-iso-octanol 19:1	0.01 M HCl, back extract 6 M NH_3	Re	[75]
trioctylamine, 0.05 M 2,2′-dichlorodiethyl ether	0.05 M benzyltrioctyl-ammonium Cl^-+0.5 M HF+8 M NH_4F	W	[76]
NaSCN, 2% (w/v)	1 M H_2SO_4, extract Et_2O	Re	[77]
thioglycolic acid, 0.4% (w/v)	0.4% (w/v) tributyl-ammonium Cl^-+0.01 M HCl, extract CH_2Cl_2	V	[78]
$Na_2S_2O_3$, 3% (w/v)	1.5 M HCl, extract MIBK, back extract 6 M NH_3+6% (w/v) H_2O_2	Fe, U, V, W	[79]
O,O,S-tripropyl-dithiophosphate, 1 M C_6H_6	7 M HCl	Ta, Ti, Th, Zn	[80]

with methyl isobutyl ketone as the organic solvent extraction is >99% complete for molybdenum if experimental conditions are closely followed. To remove the last traces of molybdenum a second extraction using thiosulfate ion as complexing agent is carried out, also with MIBK. Molybdenum forms an insoluble reddish-brown complex with thiosulfate. The MIBK fractions are combined and molybdenum back extracted into a basic aqueous phase containing hydrogen peroxide. Following acidification and removal of excess hydrogen peroxide, molybdenum is determined by any appropriate means. Cerium titration (Mo > 5 mg) or colorimetric measurement (Mo < 5 mg) are suggested by the authors. Molybdenum is separated by this procedure from Fe(III), Co(II), Ni(II), Cr(VI), Mn(II), U(VI), Ti(IV), Zr(IV), and Al(III). Tungsten and vanadium interfere through formation of tungstophosphate ion and molybdovanadophosphate ion respectively. These are extracted along with the molybdophosphate ion. Tin(II) and copper(II) intefere, the latter by forming a brown copper thiosulfate precipitate. Arsenic and excessive amounts of phosphate ion also interfere. The method is applied specifically to separation of molybdenum in various alloys and in ferromolybdenum. Following are details of the procedure.

Exactly ten mL of a sample containing from 0.6 to 5.0 mg Mo/mL [1] in 0.050 M sulfuric acid is placed in a 50 mL beaker. Add sufficient 1.00 mg P/mL, as disodium hydrogen phosphate solution, (Prepare by dissolving 1.15 g Na_2HPO_4 in distilled water and diluting to a final volume of 250 mL.) until the Mo:P ratio is 30:1 (wt/wt). For example, a solution containing 2.5 mg Mo/mL would require 0.83 mL of the phosphate solution. If Ti and/or Zr are present in the sample, a precipitate forms upon addition of phosphate. Sufficient phosphate to completely precipitate these metals is necessary before adding phosphate to establish the 30:1 Mo:P ratio. The presence of precipitate does not effect subsequent extraction. Add additional 0.050 M H_2SO_4 until the final volume is 15.0 mL. Add 1 mg hydrazine sulfate, $H_2NNH_2\cdot H_2SO_4$, for each one mg of reducible ion present. Note some knowledge of the matrix composition is necessary as ions other than molybdate are reduced by hydrazine. If chromium(VI) is present, add 5 mg hydrazine sulfate for each mg Cr(VI). Dissolve the hydrazine and gently boil the solution for 2–3 min with frequent stirring. Cool and quantitatively transfer the sample using a small volume of water to a 150 mL separatory funnel. Add sufficient 1.0 M H_2SO_4 until the H_2SO_4 concentration is 0.20 M in a volume of 20.0 mL. Dilute the same to 20.0 mL with water. Add 20.0 mL methyl isobutyl ketone and extract the sample for 2 min. Following phase separation, collect the aqueous layer in a 100 mL beaker. Reserve the MIBK solution, containing the bulk of molybdenum, for further treatment.

To the aqueous layer, add dropwise saturated bromine, $Br_2\cdot H_2O$, to destroy excess hydrazine and to return the remaining traces of molybdenum to Mo(VI). Boil the sample to remove excess bromine and transfer the solution, approximately 20 mL, quantitatively to a 150 mL separatory funnel which contains 25 mL MIBK. Add 1.0 mL 9.0 M H_2SO_4. The final aqueous phase, when diluted to 25 mL, should be 1.0 M in H^+. Add water until the aqueous volume is 23.5 mL. Add 1.5 mL 40% (w/v) sodium thiosulfate, $Na_2S_2O_3$, and immediately mix the two layers. If iron(III) is present in the sample, add an additional 2 mg $S_2O_3^{2-}$ ion for every 100 mg Fe(III) before proceeding with the extraction. Continue shaking the flask for 5 min after which the aqueous phase and any precipitate present are transferred to a second 150 mL separatory funnel which already contains 25 mL MIBK and 1.5 mL 40% (w/v) $Na_2S_2O_3$. Shake this flask for 5 min and remove the aqueous phase but leave the solid residue with the MIBK. Add 1.0 mL water to the MIBK-residue phase and shake

---

1 mL = ml = 10^{-3} l

the funnel briefly. Discard the water layer. Repeat the washing step with a second 1.0 mL water.

Combine all MIBK fractions and the reddish-brown residue in a 250 mL separatory funnel. Add 45 mL of solution 0.010 M ammonia, $NH_3 \cdot H_2O$, containing 5.0 mL 6% (v/v) hydrogen peroxide, H_2O_2. Shake the funnel for 2 min and remove the aqueous phase containing the molybdenum. Add to the MIBK 22 mL of solution 0.010 M sodium hydroxide, NaOH, containing 5.0 mL 6% (v/v) H_2O_2. Again shake the mixture for 2 min. Remove the aqueous phase and combine it with that from the previous extraction. Boil the solution for 4–5 min to remove excess H_2O_2 and filter it through a course filter paper (e.g. Whatman number 40). Rinse the residue on the paper with water. Acidify the filtrate and washings with 6 M hydrochloric acid, HCl. If reduced oxides of molybdenum are present in the sample, molybdenum blue, add $Br_2 \cdot H_2O$ dropwise to reoxidize them to Mo(VI). Boil the sample again to remove excess bromine. The determination of molybdenum can now proceed by a suitable method.

3. Ion Exchange

Because of its varied chemistry in acid and base medium, molybdenum can, under specified conditions, be retained on either a cation or an anion exchange resin. In slightly acid or neutral solution, the molybdate ion, MoO_4^{2-}, interacts with anion exchange resins; however, if the solution is too basic, OH^- ions compete for the resin exchange sites and displace molybdenum. In strong acid solutions the molybdenyl ion, MoO_2^{2+}, interacts with cation resins provided complexing agents, including the acid anion, do not combine with molybdenum to form negatively charged complexes. Molybdenum retention is affected too by the nature of the solvent and is different in mixed aqueous-organic solvents than in water solutions. All these factors coupled with the variety of strong and weak, cation and anion exchange resins available are used to advantage in separating molybdenum from other ions [81–85].

In 0.1 M HCl solution containing 1 M citric acid, for example, molybdenum as its negatively charged citrate complex is not retained on a strong acid cation resin (Dowex 50) while various other ions (Cr, Cu, Fe, Ni, Pb, V) are retained [86, 87]. With a moderately strong anion exchange resin (Anionite EDE-10) in 0.01–0.1 M HCl the molybdate ion is retained but retention decreases as the HCl concentration increases to one molar and beyond [88, 89]. Optimum retention for molybdenum on the strong base resin Anionite AV-17 is from pH one to three [90] and reaches a maximum at pH 2 if hydrofluoric acid is also present [91]. With the strong base resin Amberlite IRA-410, best retention is at pH 5 and elution of molybdenum achieved using 4 M NaOH [92]. A scheme for separating the acid insoluble sulfide qualitative group, including molybdenum, using ion exchange rather then selected sulfide precipitation employs the strong base resin Dowex 1 [93]. All ions of the acid insoluble sulfide group are retained by the column from concentrated HCl solution and subsequently eluted. Molybdenum elution is at about the middle of the group with 1 M NH_3-2 M NH_4NO_3 solvent. Specific separations of molybdenum from other ions by ion exchange are shown in Table 3-3.

Table 3.3. Separation of Molybdenum from Selected Ions by Ion Exchange

Resin	Ions Present	Condition	Eluting Solvents	Ref.
strong cation resins				
Amberlite IR 120 (H^+)	Mo, V	0.1 M HCl	Mo – 5% (w/v) $NaNH_4HPO_4$ V – 3 M HCl	[94]
Bio-Rad AG5OW-X8 (H^+)	Tc, Mo	1 M HOAc	Tc – 1 M HOAc Mo – retained	[95]
Bio-Rad AG5OW-X8 (H^+)	Mo, U, Ni, Mn	0.08 M tartaric acid + 0.02 M NH_4 tartrate	Mo – 0.08 M tartaric acid + 0.02 M ammonium tartrate, pH 2.7 U – 0.04 M tartaric acid + 0.06 M ammonium tartrate, pH 3.7 Ni – 0.15 M ammonium tartrate, pH 6.5 Mn – 2 M HCl	[96]
Bio-Rad AG5OW-X8 (H^+)	Mo, Pb	0.01 M HNO_3 + 0.25 M tartaric acid	Mo – 0.01 M HNO_3 + 0.25 M tartaric acid Pb – 0.01 M HNO_3 + 0.1 M tartaric acid	[97]
Dowex 50-X8 (H^+)	Mo, W, Nb, Ta	0.5 M H_2SO_4 + 1.5% (v/v) H_2O_2	Mo – 0.25 M H_2SO_4 + 1% (v/v) H_2O_2 W ⎤ Nb ⎟ – 0.25 M HNO_3 + 1% (v/v) H_2O_2 Ta ⎦	[98]
Dowex 50-X8 (H^+)	Mo, U	6 M HNO_3	Mo – 6 M HNO_3 + tetrahydrofuran 1:9 U – 3 M HNO_3 + tetrahydrofuran 1:4	[99]
Kationite KU-2 (H^+)	Re, Mo	0.1 M HCl + 5% (w/v) thiourea	Re – water Mo – retained	[100]
Kationite KU-2 (H^+)	Mo, Fe	0.25 M H_2SO_4 + 0.5 M HNO_3	Mo – 5% (v/v) NH_3 Fe – retained	[101]
strong anion resin				
Anionite AV-16 (F^-)	Mo, Nb	12 M HNO_3 + 3% (w/v) NaF	Mo – 12 M HNO_3 + 3% (w/v) NaF Nb – 2.5 M HCl + 0.5% (w/v) NaF	[102]
Anionite EDE-10P (Cl^-)	Mo, W	0.1 M HCl + 0.5 M HF	Mo – 1 M HCl W – 5% (v/v) NH_3 + 1 M NH_4Cl	[103]
Anionite EDE-10P (Cl^-)	Cr, Mo, W	0.5 M HF	Cr – water Mo – 1 M HCl W – 5% (v/v) NH_3 + 1 M NH_4Cl	[104]

Table 3.3 (continued)

Resin	Ions Present	Condition	Eluting Solvents	Ref.

Bio-Rad AGl-X8 (Cl$^-$)	Co, Cu, Ti, Mo	0.10 M HCl +0.05 M H$_2$C$_2$O$_4$ +0.015% (v/v) H$_2$O$_2$	Co $\vert$ – 0.10 M HCl+0.05 M
			 Cu $\vert$ H$_2$C$_2$O$_4$+0.015%
			      (v/v) H$_2$O$_2$
			 Ti – 2 M HCl+0.05 M H$_2$C$_2$O$_4$
			      +0.015% (v/v) H$_2$O$_2$
			 Mo – 0.5 M NH$_3$+1.0 M NH$_4$NO$_3$
				[105]

Bio-Rad AGl (N$_3^-$)	Mn, Mo, Re	0.5 M HN$_3$	Mn – 0.1 M HN$_3$ Mo – 0.5 M NaOH+0.5 M NaCl Re – 4 M HNO$_3$	[106]
De-Acidite FF (Cl$^-$)	Co, W, Nb, Mo	1 M HF	Co – 1 M HF W – 10 M HCl+3 M HF Nb – 7 M HCl+0.2 M HF Mo – 0.7 M HF+0.2 M NH$_4$F+4 M NH$_4$Cl	[107]
DeAcidite FF (NO$_3^-$)	V, Cr, Mo	pH 2.5, HCl	V – 0.6 M NaOH Cr – 8 M HCl Mo – 1 M HCl	[108]
Dowex 1-X2 (acetate)	Mo, W	pH 5.3, NH$_3$-NH$_4$OAc	Mo – 2 M NH$_4$OAc, pH 5.3 W – NH$_3$, pH 9	[109]
Dowex 1 (Cl$^-$)	U, W, Mo	9 M HCl+1 M HF	U – 0.5 M HCl+1 M HF W – 7 M HCl+1 M HF Mo – 1 M HCl	[110]
Dowex 1 (Cl$^-$)	Fe, Mo	9 M HCl+1 M HF	Fe – 0.01 M HCl+1 M HF Mo – 1 M HCl	[110]
Dowex 1 (Cl$^-$)	Mo, Tc	1 M HCl	Mo – 1 M HCl Tc – 4 M HNO$_3$	[111]
Dowex 1-X8 (Cl$^-$)	Mo, Re	1 M HCl	Mo – 1 M HCl Re – 2.5 M HNO$_3$	[112]
Dowex 1 (Cl$^-$)	B, Mo	pH 3, H$_2$SO$_4$	B – 0.01 M HCl Mo – 0.5 M NH$_3$+1 M KNO$_3$	[113]
Dowex 1-X8 (Cl$^-$)	V, Mo, Tl	3.6 M HCl +Br$_2$·H$_2$O trace	V – 3.6 M HCl+trace Br$_2$ Mo – 1 M HCl+2 M HClO$_4$ Tl – 9 M H$_2$SO$_4$	[114]
Dowex 1-X8 (F$^-$)	V, Ti, Mo, W	25.5 M HF	V – 10.1 M HF Ti $\vert$ – 27.5 M HF Mo$\vert$ W – 7 M HCl+3 M HF	[115]
Dowex 1-X8 (NO$_3^-$)	Mo, Re	6 M HNO$_3$+ MeOH 1:9	Mo – 6 M HNO$_3$+MeOH 1:9 Re – 9 M HCl+ tetrahydrofuran 3:7	[116]
Dowex 1-X8 (C$_2$O$_4^{2-}$)	Zr, Nb, Ta, W, Mo	1.5 M HCl +0.5 M H$_2$C$_2$O$_4$	Zr $\vert$ – 1.5 M HCl+0.5 M H$_2$C$_2$O$_4$ Nb$\vert$ +0.007 M H$_2$O$_2$ Ta – 3 M HCl+0.5 M H$_2$C$_2$O$_4$ W – 4 M HCl+0.1 M citric acid Mo – 1.9 M HCl+0.44 M NH$_4$ citrate	[117]
Dowex 1-X8 (SCN$^-$)	Re, Mo	0.5 M HCl +0.5 M NH$_4$SCN	Re – 0.5 M HCl+0.5 M NH$_4$SCN Mo – 2.5 M NH$_4$NO$_3$	[118]

Table 3.3 (continued)

Resin	Ions Present	Condition	Eluting Solvents	Ref.
Dowex 1-X8 (SCN$^-$)	Sb, Sn, Mo	3 M HCl	Sb – 0.5 M H$_2$SO$_4$ Sn ⎰– 0.5 M NaOH Mo ⎱ +0.5 M NaCl	[119]
Dowex 21K (malonate)	Mo, Cr, Fe	pH 5.6, malonic acid	Mo ⎰– retained Cr ⎱ Fe – water	[120]
weak anion resin				
Aliquat 336 (Cl$^-$) on Rohn & Haas XAD-2	V, W, Mo	0.15 M H$_2$SO$_4$	V – 0.15 M H$_2$SO$_4$+0.5% (v/v) H$_2$O$_2$ W – 0.30 M H$_2$SO$_4$+0.5% (v/v) H$_2$O$_2$ Mo – 4.0 M H$_2$SO$_4$+0.5% (v/v) H$_2$O$_2$	[121]
Amberlite CG-4B (SO$_4$)	In, U, Mo	0.1 M H$_2$SO$_4$	In – 0.5 M H$_2$SO$_4$+0.15% (v/v) H$_2$O$_2$ U – 3 M HCl+0.15% (v/v) H$_2$O$_2$ Mo – 0.5 M NaOH+0.5 M NaCl	[122]
DEAE cellulose (acetate)	U, Sm, Mo, Bi, Th	7.6 M HNO$_3$+17 M HOAc 1:9	U ⎰–7.6 M HNO$_3$+17 M HOAc Sm ⎱ Mo – 13 M HNO$_3$+17 M HOAc 1:9+0.2% (v/v) H$_2$O$_2$ Bi – 13 M HNO$_3$+MeOH 1:19 Th – 1.1 M HCl	[123]
DEAE cellulose (SCN$^-$)	Re, Mo, W	pH 3, HCl-HOAc	Re – HCl-HOAc, pH 3+0.02 M NH$_4$SCN Mo – HOAc+NaOAc, pH 5.0+0.1 M NH$_4$SCN W – 0.1 M NaOH+0.1 M NaCl	[124]
TEAE cellulose (Cl$^-$)	Zn, Mo, Pb	0.01 M HCl	Zn ⎰– 0.01 M HCl Mo ⎱ Pb – 3 M HCl	[125]

4. Paper and Thin Layer Chromatography

Paper chromatographic separation of molybdenum from interfering ions prior to appropriate quantitative measurement is often successful depending upon the proper choice of complexing agent, solvent, and mode of development [126–128]. Papers are sometimes impregnated with ion exchange particles or adsorbants, for example tin(IV) tungstate [129], to achieve a desired result. Molybdenum is separated from rhenium [130] and from chromium and tungsten [131] by paper electrophoresis. Table 3-4 lists specific separations of molybdenum using paper chromatography.

Table 3.4. Separation of Molybdenum from Selected Ions by Paper Chromatography

Paper and Method	Ions Present	Developing Solvent	R_f		Ref.
Archer 302 (ascending)	Ti, V, Mo	12 M HCl-3% (v/v) H_2O_2-H_2O-n-BuOH 1:1:8:10	Ti V Mo	0.09 0.27 0.79	[132]
FN6 paper (ascending)	V, W, Mo, Re, Fe	pentanol-12 M HCl-30% (v/v) H_2O_2 19:6:2	V W Mo Re Fe	0.11 0.45 0.72 0.99 1.00	[133]
Whatman 1 (ascending)	Cr, V, W, Mo	sym-collidine-i-BuOH-12 M HCl 10:38:52	Cr V W Mo	0.28 0.37 0.68 0.81	[134]
Whatman 3 (ascending)	Mo, Sb, Sn	MeOAc-12 M HNO_3-H_2O-sat'd Na_2H_2EDTA-1 M tartaric acid 17:1:2:3:1	Mo Sb Sn	0.38 0.72 0.83	[135]
paper, impregnate with 30% (v/v) TBP C_6H_6	W, Mo, V	5 M HNO_3	W Mo W	0.00 0.68 0.96	[136]
paper (ascending)	Mo, Re	0.1% (w/v) sali-cyclic acid 95% EtOH	Mo Re	0.00 0.38	[137]
paper	W, V, Mo, Cr	acetone-6 M HCl-1 M H_2SO_4 90:1:9	W V Mo Cr	0.00 0.39 0.44 0.88	[138]

Thin layer sorbents other than paper also separate molybdenum from other ions under proper conditions. Silicon dioxide [139–141], aluminum oxide [137, 142, 143], and specialized coatings [144–146] have all been used for this purpose. Solution conditions and choice of developing solvent effect molybdenum retention [147]. Specific examples of the thin layer technique include separation of molybdenum from technetium [148, 149] and from rhenium and tungsten [150].

5. Other Separation Techniques

Molybdenum is selectively separated from certain other elements by adsorption. Passage through columns containing aluminum oxide retain molybdate ion from sulfuric acid solutions at pH 4–6 while sulfate and other common anions pass through [151]. Activated carbon (Norit) retains rhenium from 1 M H_2SO_4 solution while molybdenum passes through the column [152]. The sorbent SG retains both molybdenum and germanium from 0.001 M H_2SO_4 solution [153]. Molybdenum is removed with 2.5 M H_2SO_4 followed by germanium with 9 M HCl. Mixtures of

Table 3.5. Separation of Molybdenum in Specific Materials

Material	Method	Reference
rocks and minerals		
silicate rocks	co-precipitation	[5]
rocks	co-precipitation	[161]
rocks	extraction	[162]
molybdenite	paper chromatography	[163]
molybdenite	column chromatography	[164]
molybdenite	co-precipitation	[11]
powellite	paper chromatography	[165]
scheelite	extraction	[50, 166]
Cu ore	extraction	[167, 168]
Mn ore	ion exchange	[114]
ore	ion exchange	[169, 170]
ore	extraction	[171, 172]
ore	column chromatography	[173]
ore	thin layer chromatography	[174]
ferrous alloys		
cast iron	extraction	[175]
cast iron	ion exchange	[176]
ferromolybdenum	extraction	[66]
Cr-Ni steel	extraction	[50, 177]
Cr-Ni steel	ion exchange	[178]
stainless steel	extraction	[53, 179]
stainless steel	column chromatography	[180]
Ni steel	column chromatography	[181]
W steel	ion exchange	[108]
low alloy steel	ion exchange	[107]
high alloy steel	extraction	[182]
steel	column chromatography	[183]
steel	thin layer chromatography	[184]
steel	extraction	[185, 186, 187, 188]
steel	ion exchange	[78, 189, 190, 191, 192]
non ferrous alloys		
Co alloy	ion exchange	[117]
Nb alloy	ion exchange	[102]
Re alloy	ion exchange	[193]
Ti alloy	extraction	[194]
Ti alloy	ion exchange	[195]
U alloy	extraction	[196]
W alloy	ion exchange	[197]
pure elements		
Ti	paper chromatography	[198]
U	precipitation	[199]
biological samples		
cow liver	paper chromatography	[200]
bone	extraction	[201]
fruits	extraction	[202]
vegetables	column chromatography	[203]

Table 3.5 (continued)

Material	Method	Reference
vegetable oils	paper chromatography	[204]
wheat	co-precipitation	[205]
plants	co-precipitation	[206]
plants	ion exchange	[207]
plants	paper chromatography	[208]
soils	paper chromatography	[209, 210]
environmental samples		
natural water	column chromatography	[211, 212]
natural water	co-precipitation	[213, 214]
sea water	co-precipitation	[215, 216, 217, 218]
sea water	column chromatography	[219]
sea water	extraction	[220, 221]
sea water	floation	[222]
sea water	ion exchange	[223, 224]
sediments	extraction	[225]
salt brines	column chromatography	[226]
flue dust	ion exchange	[227]
fly ash	ion exchange	[228]
miscellaneous samples		
fertilizier	extraction	[229, 230]
paint	thin layer chromatography	[231]

V, Mo, W, Co, and Cr, as their 8-hydroxyquinoline complexes, in chloroform are completely separated on a silica column when tetrahydrofuran-chloroform(6:4) is used with a high performance liquid chromatographic procedure [154]. Other chelating agents attached to solid supports also separate molybdenum from mixtures [155–158]. Alkali and alkaline earth elements are separated from molybdenum based upon size exclusion using a column packed with Sephadex G-10 [159]. Re, Mo, Cr, W, and V are separated at pH 6 by foam floation in the presence of hexadecyldimethylbenxylammonium chloride [160]. Selectivity is achieved by varying the chloride ion concentration.

6. Applications of Specific Materials

Table 3-5 lists techniques for the separation of molybdenum from specific materials. Details of these procedures can be found by consulting the appropriate references.

References

1. Plotnikov, V.I., Kochetkov, V.L.: Zh. Anal. Khim., *23*(3), 377 (1968)
2. Lukovnikov, A.F., Ponomarev, A.N.: Radikokhimiya, *4*, 19 (1962)

3. Lebedinskaya, M.P., Chuiko, V.T.: Zh. Anal. Khim., *28*(12), 2413 (1973)
4. Yamazaki, H., Gohda, S., Nishikawa, Y.: Bunseki Kagaku, *29*, 58 (1980)
5. Chan, K.M., Riley, J.P.: Anal. Chim. Acta, *36*, 220 (1966)
6. Plotnikov, V.I., Kochetkov, V.L.: Izv. Akad. Nauk Kaz. SSR, Ser. Fiz.-Mat., *5*(4), 53 (1967); Chem. Abstr., *69*, 90509z (1968)
7. Novikov, A.I., Kopylova, N.V.: Dokl. Akad. Nauk Tadzh. SSR, *15*(5), 38 (1972); Chem. Abstr., *80*, 64172s (1974)
8. Plotnikov, V.I., Kochetkov, V.L.: Izv. Akad. Nauk Kaz. SSR, Ser. Fiz.-Mat., *5*(6), 10 (1967); Chem. Abstr., *69*, 73616q (1968)
9. Tanaka, M.: Mikrochim. Acta, 204 (1958)
10. Novikov, A.I.: Zh. Anal. Khim., *16*(5), 588 (1961)
11. Yatirajam, V., Ahuja, U., Kakkar, L.R.: Talanta, *22*, 315 (1975)
12. Stoppel, A.E., Sidener, C.F., Brinton, P.H.M.P.: Chem. News, *130*, 353 (1925)
13. Koppel, I.: Chem.-Ztg., *48*, 801 (1924)
14. Lektorskaya, N.A., Kovalenko, P.N., Azarkhina, I.M.: Sovrem. Metody Khim. Tekhnol. Kontr. Proizvod. 151 (1968); Chem. Abstr., *72*, 8972a (1970)
15. Majumdar, A.K., Bhowal, G.: Anal. Chim. Acta, *48*, 192 (1969)
16. Simson, T.F., Rozhdestvenskaya, Z.B., Songina, O.A., Koval', A.V.: Zh. Anal. Khim., *24*(9), 1352 (1969)
17. Golubeva, I.A.: Zh. Anal. Khim., *24*(3), 467 (1969)
18. Alimarin, I.P., Polyanskĭ, V.N.: Zh. Anal. Khim., *8*(5), 266 (1953)
19. Nelidow, I., Diamond, R.M.: J. Phys. Chem., *59*, 710 (1955)
20. Yamamoto, S.: Nippon Kagaku Zasshi, *76*, 417 (1955)
21. Zelikman, A.N., Kalinina, I.G., Smol'nikova, R.A.: Zh. Neorg. Khim., *13*(10), 2778 (1968)
22. Busev, A.I., Frolkina, V.A., Koroleva, M.Ya.: Zh. Anal. Khim., *24*(2), 205 (1969)
23. Umland, F., Wvensch, G.: Fresenius' Z. Anal. Chem., *225*, 362 (1967)
24. Klitina, V.I., Sudakov, F.P., Alimarin, I.P.: Zh. Anal. Khim., *20*(11), 1145 (1965)
25. Yatirajam, V., Ram, J.: Talanta, *20*, 885 (1973)
26. Klitina, V.I., Sudakov, F.P., Alimarin, I.P.: Zh. Anal. Khim., *21*(3), 338 (1966)
27. Lakshmanan, V.I., Haldor, B.C.: Proc. Nucl. Rad. Chem. Symp., 398 (1967); Chem. Abstr., *70*, 41328b (1969)
28. Jeżowska-Trzebiatowska, B., Kopacz, S., Bartecki, A.: Zh. Neorg. Khim., *13*(7), 1899 (1968)
29. Baram, I.I.: Zh. Prikl. Khim. (Leningrad), *42*(6), 1411 (1969)
30. Busev, A.I., Frolkina, V.A.: Zh. Neorg. Khim., *9*(10), 2481 (1964)
31. De Silva, M.E.M.: Analyst (London), *100*, 517 (1975)
32. Busev, A.I., Rodionova, T.V.: Vestn. Mosk. Univ. Khim., *23*(5), 63 (1966); Chem. Abstr., *70*, 23540k (1969)
33. Busev, A.I., Rodionova, T.V.: Zh. Neorg. Khim., *14*(5), 1283 (1969)
34. Busev, A.I., Rodionova, T.V., Alvarez Escartin, L.A.: Zh. Neorg. Khim., *15*(6), 1624 (1970)
35. Laskorin, B.N., Fedorova, L.A., Egorov, I.F., Lyubosvetova, N.A.: Zh. Prikl. Khim. (Leningrad), *49*(2), 353 (1976)
36. Fischer, C., Muehl, P., Guenzler, G.: Z. Chem. *8*, 235 (1968)
37. Vieux, A.S., Rutagengwa, N., Mpeti, N.: Analusis, *4*, 134 (1976)
38. Kamiya, S., Takutomi, M., Matsuda, Y.: Bull. Chem. Soc. Jpn., *40*, 407 (1967)
39. Bantysh, A.N., Dobizha, E.V., Knyazev, D.A.: Zh. Neorg. Khim., *12*(8), 2165 (1967)
40. Rudenko, N.P., Awad, K.K., Kuznetsov, V.I., Gudym, L.S.: Vestn. Mosk. Univ., Khim., *23*(5), 36 (1968); Chem. Abstr., *70*, 14899g (1969)
41. Gorina, D.O.: Ukr, Khim. Zh., *41*(7), 773 (1975)

42. Maeck, W.J., Kussy, M.E., Rein, J.E.: Anal. Chem., *33*, 237 (1961)
43. Bantysh, A.N., Knyazev, D.A.: Zh. Neorg. Khim., *13*(1), 231 (1968)
44. Iwasaki, K., Tanaka, K., Takagi, N.: Bunseki Kagaku, *23*, 1179 (1974)
45. Wyttenbach, A., Bajo, S.: Anal. Chem., *47*, 2 (1975)
46. Ohashi, K., Shishiki, M., Yamamoto, K.: Bunseki Kagaku, *30*, 51 (1981)
47. Donaldson, E.M.: Talanta, *23*, 417 (1976)
48. Lee, A.P., Boltz, D.F.: Anal. Chim. Acta, *78*, 466 (1975)
49. Ishibashi, N., Kohara, H., Abe, K.: Bunseki Kagaku, *17*, 154 (1968)
50. Mit'kina, L.I., Mel'chakova, N.V., Peshkova, V.M.: Zh. Anal. Khim., *36*(6), 1099 (1981)
51. Agrawal, Y.K., Maru, P.C., Sharma, T.P., Patke, S., Verma, P.C.: Fresenius' Z. Anal. Chem., *276*, 300 (1975)
52. Ternero Radriguez, M.: Analyst (London), *107*, 41 (1982)
53. Arunachalam, M.K., Kumaran, M.K.: Talanta, *21*, 355 (1974)
54. Tamhina, B., Herak, M.J.: Mikrochim. Acta, I, 47 (1977)
55. Izquierdo, A., Calmet, J.: Analusis, *4*, 200 (1976)
56. Ganago, L.I., Ivanova, I.F.: Zh. Anal. Khim., *35*(6), 1138 (1980)
57. Alexandrov, A., Kostova, M.: Mikrochim. Acta, I, 487 (1976)
58. Busev, A.I., Rudzit, G.P.: Zh. Anal. Khim., *19*(5), 569 (1964)
59. Lee, Λ.P., Boltz, D.F.: Anal. Lett., *8*, 345 (1975)
60. Nacu, Al., Nacu, D., Mocanu, R.: An. Stiint. Univ. "Al. I Cuza," Iasi, Sect. Ic, *12*(1), 27 (1966); Chem. Abstr., *67*, 6192u (1967)
61. Pyatnitskii, I.V., Kravtsova, L.F.: Ukr. Khim. Zh., *35*(1), 77 (1969)
62. Kiss, A., Hegedus, A.J.: Mikrochim. Acta, 771 (1966)
63. Shustova, M.B.: Tr. Kom. Anal. Khim., Akad. Nauk SSSR, Inst. Geokhim. Anal. Khim., *15*, 111 (1965); Chem. Abstr., *63*, 4936h (1965)
64. Laskorin, B.N., Ul'yanova, V.S., Sviridova, R.A.: Zh. Prikl. Khim. 35, 2409 (1962)
65. Kletenik, Yu.B., Bykhovskaya, I.A., Sekretova, L.V.: Zh. Anal. Khim., *24*(5), 707 (1969)
66. Yatirajam, V., Ram, J.: Talanta, *21*, 439 (1974)
67. Petrov, B.I., Degtev, M.I., Zhivopistsev, V.P.: Zh. Anal. Khim., *31*(6), 1076 (1976)
68. Iordanov, N., Mareva, St., Borisov, G., Iordanov, B.: Talanta, *15*, 221 (1968)
69. Grubitsch, H., Heggeboe, T.: Monatsh. Chem., *93*, 274 (1962)
70. Yatirajam, V., Prosad, R.: Indian J. Chem., *3*, 345 (1965)
71. Yatirajam, V., Prosad, R.: Indian J. Chem., *3*, 544 (1965)
72. Kiss, A.: Magy. Kem. Foly., *69*, 524 (1963); Chem. Abstr., *61*, 85e (1964)
73. Sokolov, V.A., Levin, V.I.: Zh. Neorg. Khim., *9*(3), 742 (1964)
74. Ponomareva, A.A.: Tr. Novocherk. Politekh. Inst., (266), 103 (1972); Chem. Abstr., *78*, 131717x (1973)
75. Karag'ozov, L., Vasilev, Kh.: Hydrometallurgy, *4*, 51 (1979); Chem. Abstr., *90*, 175394s (1979)
76. Tseryuta, Yu.S., Bagreev, V.V., Arginskaya, N.A., Gushchin, N.V., Basov, A.S., Zolotov, Yu.A.: Zh. Anal. Khim., *28*(5), 946 (1973)
77. Mubayadzhyan, M.A.: Nauchn. Tr. Nauchn.-Issled. Gornometall. Inst. Yerevan, (6), 295 (1967); Chem. Abstr., *69*, 62039c (1968)
78. Ziegler, M., Horn, H.G.: Fresenius' Z. Anal. Chem., *166*, 362 (1959)
79. Yatirajam, V., Ram, J.: Mikrochim. Acta, 77 (1973)
80. Busev, A.I., Rodionova, T.V.: Anal. Lett., *2*, 9 (1969)
81. Shishkov, D.A., Shishkova, L.G.: Talanta, *12*, 857 (1965)
82. Popov, I.F., Matrenkin, V.F., Klebanov, O.B.: Sin. Svoistra Ionoobmen. (Mater Chmutov, K.V. ed.) 1968, Moscow, Izd. "Nauka"; Chem. Abstr., *71*, 82225e (1969)

83. Reingol'd B.M., Sinakevich, A.S., Khlebnikova, G.A.: Nauchn. Tr. Irkutsk. Gos. Nauchn.-Issled. Inst. Redk. Tsvetn. Met., (19), 208 (1968); Chem. Abstr. 71, 52435f (1969)

84. Cleyrergue, C., Deschamps, N., Albert, P.: Nat. Bur. Stand. (U.S.), Spec. Publ. (312), 646 (1969)

85. Shishkova, L.G., Shishkov, D.A.: Dokl. Bolg. Akad. Nauk, 22(10), 1147 (1969); Chem. Abstr., 72, 47858n (1970)

86. Klement, R.: Fresenius' Z. Anal. Chem., 136, 17 (1952)

87. Shishkov, D.A., Koleva, E.G.: C. R. Acad. Bulg. Sci., 17, 909 (1964); Chem. Abstr., 62, 13887f (1965)

88. Shishkov, D.A., Shishkova, L.G.: C. R. Acad. Bulg. Sci., 18, 235 (1965); Chem. Abstr., 63, 12356b (1965)

89. Gaibakyan, D.S., Karagezyan, A.S.: Arm. Khim. Zh. 22(11), 986 (1969); Chem. Abstr., 72, 70953y (1970)

90. Shamsiev, S.M., Senyavin, M.M.: Fiz.-Khim. Tekhnol. Issled. Miner. Syr'ya (Tashkent: Nauka), 105 (1965); Chem. Abstr., 66, 22585j (1967)

91. Eristavi, D.I., Brouchek, F.I., Macharashvili, T.G.: Tr. Gruz. Politekh. Inst., (2), 52 (1968); Chem. Abstr., 70, 73796u (1969)

92. Nomitsu, T., Fujinaka, H.: Yamaguchi Daigaku Rika, Hokoku, 10, 107 (1959); Chem. Abstr., 54, 14868e (1960)

93. Preobrazhenskii, B.K., Moskvin, L.N.: Radiokhimiya, 3, 309 (1961)

94. Matsuo, T., Iwase, A.: Bunseki Kagaku, 4, 148 (1955)

95. Jha, S.K., De Corte, F., Hoste, J.: Anal. Chim. Acta, 62, 163 (1972)

96. Strelow, F.W.E., van der Walt, T.N.: Anal. Chem., 54, 457 (1982)

97. Strelow, F.W.E., van der Walt, T.N.: Anal. Chem., 47, 2272 (1975)

98. Fritz, J.S., Dahmer, L.H.: Anal. Chem., 37, 1272 (1965)

99. Feik, F., Korkisch, J.: Mikrochim. Acta, 900 (1967)

100. Gaibakyan, D.S., Darbingan, M.V.: Izv. Akad. Nauk Arm. SSA, Khim. Nauki, 15, 321 (1962); Chem. Abstr., 58, 8395d (1963)

101. Stefkin, F.S.: Uch. Zap., Mord. Gos. Univ. 60(1), 42 (1967); Chem. Abstr., 70, 83877k (1969)

102. Gudushauri, Ts.N., Brouchek, F.I.: Soobshch. Akad. Nauk Gruz. SSR, 70(3), 609 (1973); Chem. Abstr., 79, 111361u (1973)

103. Pakholkov, V.S., Ol'khin, V.D.: Zh. Prikl. Khim., 38(6), 1235 (1965)

104. Pakholkov, V.S., Panikarovskikh, V.E.: Izv. Vyssh. Ucheb. Zaved., Khim. Khim. Tekhnol., 10(2), 168 (1967); Chem. Abstr., 67, 50044w (1967)

105. Strelow, F.W.E., Weinart, C.H.S.W., Eloff, C.: Anal. Chem., 44, 2352 (1972)

106. Oguma, K., Maruyama, T., Kuroda, R.: Anal. Chim. Acta, 74, 339 (1975)

107. Dixon, E.J., Headridge, J.B.: Analyst (London), 89, 185 (1964)

108. Hall, F.M., Bryson, A.: Anal. Chim. Acta, 24, 138 (1961)

109. Iguchi, A.: Sci. Papers Coll. Gen. Educ., Univ. Tokyo, 6, 153 (1956); Chem. Abstr., 51, 9255e (1957)

110. Kraus, K.A., Nelson, F., Moore, G.E.: J. Am. Chem. Soc., 77, 3972 (1955)

111. Huffman, E.H., Oswalt, R.L., Williams, L.A.: J. Inorg. Nucl. Chem., 3, 49 (1956)

112. Kojima, M., Okubo, T.: Tokyo Kogyo Shikensho Hokoku, 61, 372 (1966); Chem. Abstr., 66, 72181v (1967)

113. Dimitrova, N., Christova, R.: Fresenius' Z. Anal. Chem., 268, 207 (1974)

114. Korkisch J., Steffan, I., Arrhenuis, G.: Anal. Chim. Acta, 94, 237 (1977)

115. Ferraro, T.A.: Talanta, 16, 669 (1969)

116. Korkisch, J., Feik, F.: Anal. Chim. Acta, 37, 364 (1967)

117. Bandi, W.R., Buyok, E.G., Lewis, L.L., Melnick, L.M.: Anal. Chem., 33, 1275 (1961)

118. Hamaguchi, H., Kawabuchi, K., Kuroda, R.: Anal. Chem., *36*, 1654 (1964)
119. Kawabuchi, K.: Nippon Kagaku Zasshi, *87*, 262 (1966)
120. Chakravorty, M., Khopkar, S.M.: Chromatographia, *12*, 44 (1979)
121. Fritz, J.S., Topping, J.J.: Talanta, *18*, 865 (1971)
122. Kuroda, R., Oguma, K., Kono, N., Takahaski, Y.: Anal. Chim. Acta, *62*, 343 (1972)
123. Kuroda, R., Konde, T., Oguma, K.: Talanta, *20*, 533 (1973)
124. Ishida, K., Kuroda, R.: Anal. Chem., *39*, 212 (1967)
125. Kuroda, R., Oguma, K., Suzuki, M., Fuchu, Y.: Bunseki Kagaku, *24*, 1 (1975)
126. Almássy, G., Straub, J.: Magy. Kem. Foly., *60*, 104 (1954); Chem. Abstr., *52*, 6053a (1958)
127. Termendzhyan, Z.Z., Gaibakyan, D.S.: Arm. Khim. Zh., *27*, 25 (1974); Chem. Abstr., *81*, 57816x (1974)
128. Mao, C.-H., Chen., P.: Hua Hsueh Hsueh Pao, *37*, 71 (1979); Chem. Abstr., *91*, 67839c (1979)
129. Qureshi, M., Akhtar, I., Mathur, K.N.: Anal. Chem., *39*, 1766 (1967)
130. Yin, P.-H., Chen, T.T.: Hua Hsueh Tung Pao, 375 (1964); Chem. Abstr., *62*, 2229c (1965)
131. Blum, L.: Rev. Chim. (Bucharest), *9*, 28 (1958)
132. Lederer, M.: Anal. Chim. Acta, *8*, 259 (1953)
133. Geyer, R., Hermann W., Bogatzki, B.: Wiss. Z. Tech. Hochsch. Chem. "Carl Schorlemner," Leuna-Mersburg, *9*, 1 (1967); Chem. Abstr., *67*, 121950b (1967)
134. Witwit, A.S., Magee, R.J., Wilson, C.L.: Talanta, *9*, 495 (1962)
135. Gallego Andreu, R., Bernal, J.L.: An. Quim., *73*, 683 (1977)
136. Hu, Z.-T., Shi, S.-C.: Hua Hsueh Tung Pao, 312 (1963); Chem. Abstr., *60*, 6d (1964)
137. Gaibekyan, D.S., Sarkisyan, Zh.V.: Zh. Anal. Khim., *29*, 1991 (1974)
138. Isaeva, K.G., Viktorova, M.E.: Zh. Anal. Khim., *29*, 382 (1974)
139. Johri, K.N., Kaushik, N.K., Singh, K.: Mikrochim. Acta, 737 (1962)
140. Beneš, J.: Collect. Czech. Chem. Commun., *44*, 1406 (1979)
141. Qureshi, M., Sethi, B.M., Sharma, S.D.: Sep. Sci. Tech., *15*, 1685 (1981)
142. Gaibakyan, D.S.: Arm. Khim. Zh., *22*, 13 (1969); Chem. Abstr., *71*, 27107t (1969)
143. Shiobara, Y.: Bunseki Kagaku, *19*, 243 (1970)
144. Renault, N., Deschamps, N.: Radiochem. Radioanal. Lett., *13*, 207 (1973)
145. Kitaeva, L.P., Volynets, M.P., Suvorova, S.N.: Zh. Anal. Khim., *34*, 922 (1979)
146. Kuroda, R., Hosoi, N.: Chromatographia, *14*, 359 (1981)
147. Kitaeva, L.P., Volynets, M.P., Suvorova, S.N.: Zh. Anal. Khim., *35*, 301 (1980)
148. Maki, Y., Tanaka, K., Murakami, Y.: Kanagawa-Ken Kogyo Shikensho Kenkyu Hokoku, (20), 40 (1968); Chem. Abstr., *70*, 16309p (1969)
149. Mikulaj, V., Dobias, M.: Acta Fac. Rerum Nat. Univ. Comenianae, Chim., (12), 125 (1968); Chem. Abstr., *71*, 44714f (1969)
150. Kuroda, R., Kawabuchi, K., Ito, T.: Talanta, *15*, 1486 (1968)
151. Kühn, K.: Fresenius' Z. Anal. Chem., *130*, 210 (1949)
152. Alexander, G.B.: J. Am. Chem. Soc., *71*, 3043 (1949)
153. Akerman, K., Wiater, D., Kozak, Z., Jurkiewicz, K., Krol, T.: Przem. Chem., *43*, 442 (1964); Chem. Abstr., *61*, 12986g (1964)
154. Wenclawiak, B.: Fresenius' Z. Anal. Chem., *310*, 144 (1982)
155. Fritz, J.S., Dahmer, L.H.: Anal. Chem., *40*, 20 (1968)
156. D'Olieslager, W., Indesteege, J., D'Hont, M.: Talanta, *22*, 395 (1975)
157. King, J.N.: Nat. Tech. Inform. Service (U.S.), Report No. IS-T-801, 1978
158. Moroshkina, T.M., Serbina, A.M., Limonova, L.N.: Vestn. Leningr. Univ., Fiz. Khim, (1), 93 (1979); Chem. Abstr., *91*, 198513q (1979)
159. Karajannis, S., Ortner, H.M., Spitzy, H.: Talanta, *19*, 903 (1972)

160. Charewicz, W., Grieves, R.B.: Anal. Lett. *7*, 233 (1974)
161. Kuznetsov, V.I., Myasoedova, G.A.: Tr. Kom. Anal. Khim. Akad. Nauk SSSR, Inst. Geokhim. Anal. Khim., *9*, 89 (1958); Chem. Abstr., *53*, 3981b (1959)
162. Gil'bert, E.N., Trunova, V.A., Torgov, V.G., Bobrov, V.A., Olrazovskii, E.G.: Zavod. Lab., *47*(5), 6 (1981)
163. Duca, Al., Stanescu, D., Puscasu, M.: Rev. Roum. Chim., *11*, 839 (1966); Chem. Abstr., *65*, 19290a (1966)
164. Duca, Al. Stanescu, D., Puscasu, M.: Rev. Roum. Chim., *11*, 833 (1966); Chem. Abstr., *66*, 25702f (1967)
165. Agrinier, H.: C. R. Acad. Sci., Paris, *249*, 2365 (1959)
166. Busev, A.I., Rodionova, T.V.: Zh. Anal. Khim., *23*(6), 877 (1968)
167. Caiozzi, M., Zunino, H., Sepulveda, L.: Talanta, *16*, 1590 (1969)
168. Bustos, L.I.: Analusis, *6*, 75 (1978)
169. Shishkov, D.A., Shishkova, L.G.: Khim. Ind. (Sofia), *35*, 210 (1963); Chem. Abstr., *61*, 2454f (1964)
170. Gaibakyan, D.S., Darbinyan, W.V.: Izv. Akad. Nauk Arm. SSR, Khim. Nauki, *17*, 631 (1964); Chem. Abstr., *63*, 2418h (1965)
171. Skobeev, I.K., Sinakevich, A.S., Bavik, N.V.: Tsvekn. Metall., *42*(9), 77 (1969); Chem. Abstr., *72*, 34489t (1970)
172. Ise, K.: Bunseki Kagaku, *30*, 629 (1981)
173. Brandone, A., Meloni, S., Girardi, F., Sabbioni, E.: Analusis, *2*, 300 (1973)
174. Sen, A.K., Ghosh, U.C.: J. Liq. Chromatogr., *3*, 71 (1980)
175. Thompson, B.A., La Fleur, P.D.: Anal. Chem., *41*, 852 (1969)
176. Lanza, P., Ferri, D., Buldini, P.L.: Analyst (London), *105*, 379 (1980)
177. Bhowal, S.K., Umland, F.: Fresenius' Z. Anal. Chem., *282*, 197 (1976)
178. Shemyakin, F.M., Kharlamov, P.P., Mitselovskii, E.S.: Zavod. Lab., *16*, 1126 (1950)
179. Dhara, S.C., Khopkar, S.M.: Indian J. Chem., *5*, 12 (1967)
180. Fritz, J.S., Beuerman, D.R.: Anal. Chem., *44*, 692 (1972)
181. Fritz, J.S., Hedrick, C.E.: Anal. Chem., *36*, 1324 (1964)
182. McKaveney, J.P., Freiser, H.: Anal. Chem., *29*, 290 (1957)
183. Lin, J.: Fen Hsi Hua Hsueh, *8*, 345 (1980); Anal. Abstr., *42*, 3B234 (1982)
184. Rai, J., Kukreja, V.P.: Chromatographia, *2*, 404 (1969)
185. Alimarin, I.P., Medvedeva, A.M.: Tr. Kom. Anal. Khim. Akad. Nauk SSSR, Inst. Geokhim. Anal. Khim., *6*, 351 (1955); Chem. Abstr., *50*, 12741a (1956)
186. Hall, F.M., Bryson, A.: Anal. Chim. Acta, *24*, 138 (1961)
187. Stepin, V.V., Ponosov, V.I.: Tr. Ural. Nauchno-Issled. Inst. Chern. Met., *2*, 272 (1963); Chem. Abstr., *61*, 3665g (1964)
188. Busev, A.I., Rodionova, T.V.: Anal. Lett., *3*, 235 (1970)
189. Young, R.S., Leibowitz, A.: Iron Age, *164*(21), 75 (1949)
190. Alimarin, I.P., Medvedeva, A.M.: Zavod. Lab., *21*, 1416 (1955)
191. Sudo, E.: Sci. Rep. Res. Inst., Tohoku Univ., Ser. A., *8*, 380 (1956); Chem. Abstr., *51*, 4201d (1957)
192. Maklakova, V.P., Ryazanov, I.P.: Zavod. Lab., *34*, 1049 (1968)
193. Ryabchikov, D.I., Borisova, L.V., Gerlit, Yu.B.: Zh. Anal. Khim., *17*(7), 890 (1962)
194. Yatirajam, V., Ram, J.: Anal. Chim. Acta, *59*, 381 (1972)
195. Kharlamov, I.P., Yakovlev, P.Ya.: Zavod. Lab., *23*, 535 (1957)
196. Fritze, K.: Radiochim. Acta, *5*, 164 (1966)
197. Wilkins, D.H.: Talanta, *2*, 355 (1959)
198. Kolier, I., Ribaudo, C.: Anal. Chem., *26*, 1546 (1954)
199. Heres, A.: CEA, R-4433 (1973); Anal. Abstr., *26*, 1472 (1974)
200. Erkelens, P.C. van: Anal. Chim. Acta, *25*, 226 (1961)

201. Healy, W.B., McCabe, W.J.: Anal. Chem., *35*, 2117 (1963)
202. Mirza, M.Y., Nwabue, F.I.: Talanta, *28*, 49 (1981)
203. Kielczewski, W., Uchman, W.: Rocz. Wyzsz. Szk. Roln. Poznaniu, *30*, 145 (1966); Chem. Abstr., *72*, 74436e (1970)
204. Gorbach, G.: Metal Catal. Lipid Oxid., SIK Symp., Pap. Discuss. Marcuse, R. editor 1967 (Pub. 1968) Goteborg, Sweden, Sv. Inst. Konserveringsforsk.; Chem. Abstr., *71*, 45369j (1969)
205. Chelnokova, M.N., Busev, A.I., Kosareva, T.M.: Zh. Anal. Khim., *36*(2), 230 (1981)
206. Orlova, L.P.: Agrokhimiya, 128 (1967); Chem. Abstr., *68*, 27412b (1968)
207. Sharrer, K., Höfner, W.: Z. Pflanzenernähr. Düng. Bodenkd., *86*, 49 (1959); Chem. Abstr., *55*, 11729c (1961)
208. Bönig, G.: Landwirtsch. Forsch., *9*, 97 (1956)
209. Bönig, G.: Landwirtsch. Forsch., *9*, 101 (1956)
210. Szilagyi, M.: ATOMKI (Atom. Kut. Inst.) Kozlem., *8*(2), 97 (1966); Chem. Abstr., *66*, 51864e (1967)
211. Korganova, T.S., Polyakov, V.A., Kolotov, B.A., Nechaeva, T.P.: Zavod. Lab., *39*, 1186 (1973)
212. Zaguzin, V.P., Ksenzova, V.I., Pogrebyak, Yu.F.: Zh. Anal. Khim., *35*(6), 1143 (1980)
213. Kuznetsov, V.I., Loginova, L.G., Myasoedova, G.V.: Zh. Anal. Khim., *13*, 453 (1958)
214. Sugawara, K., Tanaka, M.: Bull. Chem. Soc. Jpn., *32*, 221 (1959)
215. Ishibashi, M., Fujinaga, T., Kuwamoto, T.: Nippon Kagaku Zasshi, *79*, 1496 (1958)
216. Nazarenko, V.A., Grekova, I.M.: Zh. Anal. Khim., *31*(11), 2137 (1976)
217. Kim, Y.S., Zeitlin, H.: Anal. Chim. Acta., *46*, 1 (1969)
218. Kim, Y.S., Zeitlin, H.: Anal. Chim. Acta, *51*, 516 (1970)
219. Sloot, H.A.v.d., Walls, G.D., Das, H.A.: Anal. Chim. Acta, *90*, 193 (1977)
220. McLeod, C.W., Otsuki, A., Okamoto, K., Haraguchi, H., Fuwa, K.: Analyst (London), *106*, 419 (1981)
221. Pokhol'chuk, S.F., Andrianov, A.M.: Zh. Anal. Khim., *34*(1), 193 (1979)
222. Kim, Y.S., Zeitlin, H.: Sep. Sci., *6*, 505 (1971)
223. Riley, J.P., Taylor, D.: Anal. Chim. Acta, *41*, 175 (1968)
224. Kawabuchi, K., Kuroda, R.: Anal. Chim. Acta, *46*, 23 (1969)
225. Rancicova, M., Cuta, J., Malat, M.: Vodni Hospod., *31B*, 19 (1981); Anal. Abstr., *42*, 5B104 (1982)
226. Leyden, D.E., Steele, M.L., Jablonski, B.B., Somoano, R.B.: Anal. Chim. Acta, *100*, 545 (1978)
227. Meloche, V.W., Preuss, A.F.: Anal. Chem., *26*, 1911 (1954)
228. Gladney, E.S.: Anal. Lett., *11A*, 429 (1978)
229. Marsh, S.F.: Anal. Chem., *39*, 696 (1967)
230. Angers, D.W., Swanson, R.: U.S. Patent 3, 415, 616 (1968); Chem. Abstr., *70*, 39946h (1969)
231. Linde, H.G.: J. Forensic Sci., *25*, 870 (1980)

Chapter 4

Gravimetric Methods

Molybdenum(VI) precipitates with a variety of reagents, both inorganic and organic, under one of two conditions. In weak acid, neutral, or basic solution, generally pH > 4 or 5, molybdenum exists as the molybdate ion, MoO_4^{2-}. Insoluble molybdate salts form with many metal ions. Lead molybdate formation in particular is a common analytical procedure. In moderately strong acid solution, pH $\sim$ 1, the molybdenyl ion, MoO_2^{2+} predominates and precipitates with various anions. Many organic anions, especially 8-hydroxyquinoline and α-benzoinoxime, form insoluble compounds with molybdenyl ion. An insoluble hydrous molybdenum (VI) oxide, $MoO_3 \cdot xH_2O$, exists but is generally not used for quantitative work because of the tendency to redissolve either as molybdenyl ion, should the solution become too acid, or molybdate ion, should the solution become too basic. Within the approximate pH range 1–2 various isopoly molybdenum species, $Mo_xO_y^{n-}$ where x, y, and n are integers, exist. These further complicate hydrous oxide precipitation. Heteropoly molybdenum species, $Mo_xA_yO_z^{n-}$ where x, y, z, and n are integers, are also common and widely used for quantitative determination of the heteroatom, especially when A equals phosphorus. Heteropoly molybdates are generally not, however, utilized for determination of molybdenum.

1. Precipitation with Sulfide Ion

Molybdenum(VI) is precipitated as the insoluble sulfide, MoS_3, from acid solution with H_2S. Following filtration through paper the precipitate is ignited to the oxide, MoO_3, and weighed [1, 2]. The procedure is questionable in several respects. First it is not selective, other heavy metal sulfides, if present, also precipitate. It is not quantitative as various soluble thiomolybdates can form and remain in solution [3]. It is not convenient as anyone knows who has worked with sulfide precipitates. Finally, one must be cautious when igniting the sulfide precipitate to the oxide as the latter is volatile and loss of molybdenum can occur. Conversion of sulfide to oxide is generally carried out below 600 °C in an electric furnace [4] although temperatures of 425 °C [5] and 300 °C [6] are proposed to assure complete retention of molybdenum. The latter temperature is recommended for minute amounts of molybdenum where even the slightest loss represents an appreciable fraction of the total amount of molybdenum present. In all cases, ignition is continued until the crucible and its contents are brought to constant weight.

Several modifications of the sulfide precipitation procedure are available to improve its desirability. Precipitation in the presence of 5% formic acid is claimed to prevent colloidal molybdenum sulfide formation [7]. Precipitation from homogeneous solution, with internal generation of sulfide from thermal decomposition of thioacetamide, improves the physical character of the precipitate allowing for greater ease in filtering. The sample in one molar mineral acid and containing approximately one gram of dissolved thioacetamide is heated to boiling for one hour, preferably under a slight external pressure to prevent loss of generated H_2S [8, 9]. Addition of tartaric acid to the sample minimizes coprecipitation of tungsten sulfide if both are present in solution.

Regardless of the precautions taken, a second precipitation is desirable. The mother liqueur from the first filtration is boiled to remove H_2S. A few drops of bromine water, $Br_2 \cdot H_2O$, added and the solution boiled again. This oxidizes any molybdenum present which may have been reduced through reaction with sulfide ion. The solution is evaporated to a convenient volume and H_2S again added to repeat the precipitation. The initial precipitate, or combined precipitate if a double precipitation was carried out, is, after conversion to the oxide, redissolved in concentrated ammonia. Any residue which remains, heavy metal ions adsorbed unto the molybdenum sulfide precipitate, is filtered, washed, weighed, and a correction made in the weight of MoO_3 initially obtained.

Many of the problems associated with sulfide precipitation of molybdenum are diminished if molybdenum(V) rather than molybdenum(VI) sulfide is formed [10]. The molybdenum containing sample is treated with hydrazine prior to addition of hydrogen sulfide. The authors [10] report contamination of less than one percent for W(VI), V(V), Fe(III), Co(II), Ni(II), and U(VI). In addition, the presence of EDTA prevents molybdenum(V) sulfide precipitation while Cu(II), Bi(III), As(III), Sb(III), Sn(II), Pt(IV), and Pd(II) are brought down as their insoluble sulfides [10]. Some molybdenum is, however, absorbed upon these extraneous metal sulfides, for example up to one percent Mo coprecipitates with copper when the copper is present in a seven fold excess. A reprecipitation of these extraneous metals is helpful. This is followed by combining the various filtrates, removing the EDTA, condensing the solution to a suitable volume, adjusting the pH, and precipitating the molybdenum contained within the filtrates. Separation of molybdenum from lead and cadmium by precipitation as molybdenum(V) sulfide is incomplete regardless of the precautions taken [10]. Following precipitation of molybdenum(V) sulfide the solid is dried at $100\,°C$ and weighed $Mo_2S_5 \cdot 3H_2O$.

To precipitate molybdenum as molybdenum(V) sulfide according to the procedure of Yatirajam, Ahuja, and Kakkar [10] proceed as follows:

The sample should contain between 5 and 100 mg Mo. Add 12 M hydrochloric acid, HCl, dropwise until the sample is 2 M in acid. Dilute to about 10 ml with water. Add 200 mg solid hydrazine sulfate, $H_2NNH_2 \cdot H_2SO_4$. Boil for 3 to 4 min in a covered beaker. Cool and dilute to 200 ml with water. The sample is now 0.1 M in H^+. Heat the solution to $90\,°C$ and pass hydrogen sulfide gas, H_2S, through the sample rapidly for 15 min. Return the sample to $90\,°C$ and filter it hot through a previously weighed number 4 (very fine) sintered glass crucible. Wash the precipitate with four 5 ml portions of 0.1 M HCl which have been saturated with H_2S. The filtrate is collected, boiled to remove H_2S, and treated with 5 ml of 6% (v/v) hydrogen peroxide, H_2O_2, solution. Following this, evaporate the sample to about 10 ml and repeat the precipitation procedure starting with the addition of hydrazine sulfate. Filter the precipitate from this second step through the same crucible as before, thus combining all the molybdenum together. The crucible and its contents are dried at $100 \pm 2\,°C$

under vacuum for one hour, cooled in a dry (oxygen free) carbon dioxide atmosphere for 15 min, and weighed. The precipitate is $Mo_2S_5 \cdot 3H_2O$. Conversion factor, $Mo/Mo_2S_5 \cdot 3H_2O$, is 0.475.

If prior precipitation of other metals is desired, proceed as described with the hydrazine sulfate reduction. Following dilution to 200 ml add 150 mg of solid EDTA, as the disodium salt $Na_2H_2EDTA \cdot 2H_2O$. Heat the solution to 90 °C and continue as before except the extraneous metal sulfides are filtered through medium porosity ashless filter paper (e.g. Whatman number 40) rather than through a sintered glass crucible. The sulfide precipitate may contain in addition to interfering metals, some molybdenum. Wash the precipitate with several 5 mL portions of 0.1 M HCl previously saturated with H_2S. Place the folded paper containing the precipitate in a porcelain crucible, dry to remove moisture, and ignite at 500 °C for one-half hour or longer until constant weight is achieved. The metal oxide residue is dissolved by adding 5 mL of 6 M ammonia, $NH_3 \cdot H_2O$, and stirring vigorously. Molybdenum oxide dissolves most other metal oxides do not. Filter through a course filter paper (e.g. Whatman number 41), wash the precipitate with 1% (v/v) ammonia, and transfer the filtrate and washings back to the mother liqueur from which the extraneous metal sulfides originated. The combined filtrates are made acid with 12 M HCl and carefully evaporated. When the volume reaches about 20 mL add 5 mL of 18 M sulfuric acid, H_2SO_4. Continue to evaporate until dense white fumes of sulfur trioxide, SO_3, are observed coming from the sample. Dissolve the residue in a minimum amount of water. Add 5 mL of 6% (v/v) hydrogen peroxide solution and boil gently for a few minutes. The sample is now ready to begin the sulfide precipitation of molybdenum without the interference of those metals previously removed when EDTA was present in the solution.

2. Precipitation as Molybdate Ion

2.1 Lead Molybdate

Lead(II) forms an insoluble compound with molybdate ion in weak acid solution. Provided other ions which might interfere are absent, precipitation with lead is a convenient means for determining molybdenum [11]. One should avoid a large excess of lead [12] and it is neccessary to buffer the solution to assure proper pH. Lead molybdate is partially soluble if the solution is too acid and lead or other metal hydroxides precipitate if the solution is too basic. Ions which precipitate with lead, sulfate, chromate, tungstate, etc. interfere. Interfering also is iron(III) and other heavy metal ions which form insoluble hydrous oxides even in slightly acid solution. A modification of the usual procedure is to release lead ion internally within the sample, homogeneous precipitation, by displacement of Pb(II) from its EDTA complex upon addition of Cr(III) [13].

The following procedure is adapted from review sources on the determination of molybdenum [14, 15].

The sample should contain about 0.15 to 1.5 mg Mo/mL. Transfer about 50 mL of sample to a 150 mL beaker. Add 6 M HCl or 1 M NaOH dropwise as necessary to adjust the acidity to pH 4. Add 20 mL of acetic acid-ammonium acetate buffer solution (containing 5 mL glacial acetic acid, $HC_2H_3O_2$, and 15 g ammonium acetate, $NH_4C_2H_3O_2$, per 100 mL). Heat the sample to boiling and with constant stirring add dropwise from a buret 4% (w/v) lead nitrate solution, $Pb(NO_3)_2$. Addition is continued until no further precipitate formation is observed when Pb(II) comes in contact with sample. Stop the stirrer, allow the precipitate to settle, and again add a drop or two of Pb(II) solution. This addition is con-

Table 4.1. Precipitation of Metal Molybdates for Determination of Molybdenum

Metal	Condition	Drying	Weigh	Reference
Ag^+	pH 7.0–7.4	dry 250 °C	Ag_2MoO_4	[16, 17]
Ag^+	pH 7 0–7.4+0.025 M NH_4NO_3	dry 175 °C	Ag_2MoO_4	[18]
Ba^{2+}	pH 6–7	ignite 500 °C	$BaMoO_4$	[19, 20]
Hg_2^{2+}	–	ignite 500 °C	MoO_3	[21]
Sr^{2+}	pH 6–8, boil	ignite	$SrMoO_4$	[22]

tinued until one is certain no more precipitate forms. Gently boil the solution for 5 min to congeal the precipitate, then filter through a medium fine ashless paper (Whatman number 540). Wash the precipitate several times with hot 20 mL portions of 2% (w/v) ammonium nitrate solution or until all excess of Pb(II) is removed. The washings are tested by adding a drop of 1% (w/v) potassium chromate, K_2CrO_4, solution. Formation of a yellow solid indicates the presence of Pb(II) ion. Transfer the paper with precipitate to a porcelain crucible and dry at 110 °C to remove moisture. Then ignite the crucible at 600 °C until constant weight is achieved. The precipitate is $PbMoO_4$. Conversion factor, $Mo/PbMoO_4$, is 0.2613.

2.2 Other Molybdates

Many metal ions that form insoluble molybdates are used for quantitative molybdenum determination. Some of these are listed in Table 4-1.

3. Precipitation with Organic Reagents

3.1 8-Hydroxyquinoline

8-Hydroxyquinoline (8-quinolinol, oxine) precipitates many metals. Its use is made selective by proper pH control and addition of appropriate complexing agents to mask unwanted ions which otherwise might interfere with the desired metal precipitate. Quantitative precipitation of molybdenum occurs in the pH range 3.3–7.6 [23]. A pH of about 5, achieved with acetic acid-ammonium acetate buffer is generally employed. If EDTA is also present in the sample solution, interference from Pb, Bi, Hg(II), Cd, Cu(II), Fe(III), Al, Cr(III), Be, Co, Zn, Mn(II), and Ni is prevented [24]. Titanium(IV), vanadium(V), and tungsten(VI) interfere even in the presence of EDTA [24]. Molybdenum is separated from tungsten by 8-hydroxyquinoline precipitation if ammonium tartrate is added to the sample, tartrate complexing tungsten and preventing its interference [25]. Titanium is removed prior to molybdenum precipitation as the insoluble hydrous oxide [24]. Vanadium interference is eliminated by selectively dissolving the vanadium(V) oxine precipitate with 2 M HCl. Molybdenum oxine remains as a precipitate [26]. The yellow molybdenum 8-hydroxyquinoline compound, $MoO_2(C_9H_6ON)_2$ is stoichiometric and can be dried at temperatures below 138 °C without decomposing [27].

The procedure of Přibil and Malát for precipitation of molybdenum(VI) with 8-hydroxyquinoline is as follows [24]:

The sample should contain from 25 to 100 mg Mo in approximately 50 mL of solution. Add dropwise 1 M sulfuric acid, H_2SO_4, or 1 M sodium hydroxide, NaOH solution until the pH of the sample is 5. Add 15 mL of 5% (w/v) EDTA solution, $Na_2H_2EDTA \cdot 2H_2O$, followed by 5 mL of acetic acid-ammonium acetate buffer (containing 15 mL glacial acetic acid, $HC_2H_3O_2$, and 30 g ammonium acetate, $NH_4C_2H_3O_2$, per 100 mL of solution). Dilute to approximately 80 mL with distilled water and apply heat until the solution is gently boiling. Add slowly with constant stirring 3% (w/v) 8-hydroxyquinoline (Prepare by dissolving 3 g 8-hydroxyquinoline in 5 mL of glacial acetic acid, $HC_2H_3O_2$. Add 80 mL distilled water followed by dropwise addition of 4 M ammonia, NH_3, solution until cloudiness appears. Now add 4 M $HC_2H_3O_2$ until the solution is clear. Dilute to 100 mL total volume with water.) Continue reagent addition until no more precipitate formation is observed, then add additional reagent until the supernate is slightly yellow in color. Digest the sample at boiling temperature for 2–3 min. Filter through a fine (number 3) sintered glass crucible and wash several times with hot 20 mL portions of distilled water. The washings should be colorless indicating no 8-hydroxyquinoline is present. Dry at 130–135 °C for one hour and then to constant weight. The precipitate is $MoO_2(C_9H_6ON)_2$ molecular weight 416.2. Conversion factor $Mo/MoO_2(C_9H_6ON)_2$, is 0.2305.

3.2 α-Benzoinoxime

Like 8-hydroxyquinoline α-benzoinoxime (cupron) forms precipitates with several metal ions. Molybdenum is precipitated by α-benzoinoxime from acid solution [28, 29]. Many metal ions do not interfere. Interference does occur from Ta, Nb, and Si which form insoluble hydrous oxides. Chromium and vanadium if reduced to lower oxidation states do not interfere. Iron, if precipitated with molybdenum, is removed by washing the precipitate with dilute sulfuric acid [29]. Palladium and tungsten must be absent although estimation of the amount of tungsten and some other metals coprecipitated with molybdenum is possible if the precipitate is ignited to the oxide rather than weighed directly. Molybdenum from the oxide residue is redissolved and the weight of the remaining residue used to correct the total weight of precipitate obtained [15]. Addition of a mild oxidizing agent during the course of the precipitation is necessary to prevent reduction of molybdenum(VI) by α-benzoinoxime rather than precipitation [29]. It is possible to either dry the precipitated $MoO_2(C_{14}H_{12}O_2N)_2$ to constant weight at 105 °C or to ignite the precipitate to MoO_3 at 530–550 °C [30]. A homogeneous precipitation method in which the precipitating agent is formed from benzoin and hydroxylamine in the presence of molybdenum is available [31].

The following procedure is that of Hoenes and Stone [29].

To 10 mL of sample containing between 8 and 20 mg Mo add 10 mL of 10% (w/v) sulfuric acid, H_2SO_4. It is necessary that the sample contain 5% sulfuric acid before proceeding further. Add 5 mL or more of 2% (w/v) iron(II) ammonium sulfate, $FeSO_4 \cdot (NH_4)_2SO_4$, as needed to reduce Cr(VI) and/or vanadium(V) if either is present in the sample. Cool the solution in an ice bath and slowly add with stirring 10 mL of 1% (w/v) α-benzoinoxime (in 1:1 water acetone). Add dropwise saturated aqueous bromine water, $Br_2 \cdot H_2O$, until the sample is slightly colored. Allow the sample to stand for 10 min, then filter the cold solution through a course (number one) scintered glass crucible. When most of the mother liqueur has passed through the filter, add 5 mL of 1% (w/v) sulfuric acid to the precipitate as a wash solution. Transfer the precipitate to the crucible with water, being careful that the precipitate is always moist; otherwise it becomes difficult to work with. Wash the

Table 4.2. Organic Precipitating Agents for Molybdenum

Reagent	Condition	Weighing Form	Ref.
N^1-(4-chlorophenyl)-N^2-(4-chloro-o-tolyl)-N^1-hydroxybenzamidine	pH 5 acetate	ignite to MoO_3	[32]
2-methoxy-N-o-tolyl-benzohydroxamic acid	1 M HCl, 60 °C	dry 120 °C $MoO_2(C_{15}H_{14}O_3N)_2$	[33]
salicylhydroxamic acid	pH 1–2	dry 110 °C $MoO_2(C_7H_6O_3N)_2$	[34]
thiocyanate + antipyrine	0.1 M HCl	ignite to MoO_3	[35]
thiocyanate + methylene blue	1 M HCl	ignite to MoO_3	[36]
thiocyanate + tetra-ethylammonium Cl^-	3 M HCl, ascorbic acid	ignite to MoO_3	[37]
N-o-toluoyl-N-o-tolylhydroxylamine	2 M H_2SO_4	dry 110 °C $MoO_2(C_{15}H_{14}O_2N)_2$	[38]

Table 4.3. Gravimetric Determination of Molybdenum in Specific Materials

Material	Precipitation Agent	Ref.
molybdenum ore		
molybdenite	sulfide	[39, 40]
molybdenite	$[Cr(NH_3)_5Cl]^{2+}$	[41]
scheelite	8-hydroxyquinoline	[42]
ferrous alloys		
ferromolybdenum	sulfide	[43, 44]
steel	β-naphthoquinoline	[45]
Cr-Mo steel	α-benzoinoxime	[29]
steel	2-methoxy-N-o-tolylbenzohydroxamic acid	[33]
stainless steel	8-hydroxyquinoline	[46]
stainless steel	N-o-toluoyl-N-o-tolylhydroxylamine	[38]
tungsten steel	sulfide	[47]
nonferrous alloys		
cobalt based alloy	8-hydroxyquinoline	[48]
titanium based alloy	sulfide	[49]
titanium based alloy	8-hydroxyquinoline	[50]
uranium based alloy	α-benzoinoxime	[51]
molybdenum containing catalyst	8-hydroxyquinoline	[52]

precipitate thoroughly with two 5 mL portions of 1:1 water acetone mixture. Dry at 105 °C for one hour and then at half hour periods until constant weight is achieved. The precipitate is $MoO_2(C_{14}H_{12}O_2N)_2$ molecular weight 580.5. Conversion factor, $Mo/MoO_2(C_{14}H_{12}O_2N)_2$, is 0.1653.

3.3 Other Organic Precipitating Agents

Most organic precipitating agents for molybdenum lack selectivity. In addition, the molybdenum precipitate frequently cannot be brought to constant weight and must be ignited to MoO_3, an additional time consuming step. With proper pH control and addition of appropriate masking agents, these reagents do, however, successfully separate molybdenum quantitatively from rather complex matrices. See Tables 4-2 and 4-3 for further details.

4. Applications to Specific Materials

Prior to availability of modern instrumental procedures, gravimetric methods were extensively used for determination of molybdenum in a variety of materials. In instances where sufficient molybdenum is present, they are still applicable, especially if sophisticated instrumentation is not readily available. Provided suitable experimental conditions can be maintained, pH masking agents and/or prior removal of interferences, drying temperature, etc., gravimetric procedures yield accurate and precise results. Table 4-3 summarizes some applications of specific precipitating agents to various molybdenum containing materials.

References

1. Binder, O.: Chem. Ztg., *42*, 255 (1918)
2. Taimni, I.K., Agarwal, R.P.: Anal. Chim. Acta, *9*, 203 (1953)
3. Tridot, G., Bernard, J.C.: Acta Chim. Acad. Sci. Hung., *34*, 179 (1962)
4. Brinton, P.H., Stoppel, A.E.: J. Amer. Chem. Soc., *46*, 2454 (1924)
5. Wolf, K.: Z. Angew. Chem., *31*, I, 140 (1918)
6. Pavelka, F., Zucchelli, A.: Mikrochem. Ver. Mikrochim. Acta, *31*, 69 (1943)
7. Sterba-Böhm, J., Vostrebal, J.: Z. Anorg. Allg. Chem., *110*, 81 (1920)
8. Flaschka, H., Jakobljevich, H.: Anal. Chim. Acta, *4*, 306 (1950)
9. Burriel-Marti, F., Vidan, A.M.: Anal. Chim. Acta, *26*, 163 (1962)
10. Yatirajam, V., Ahuja, U., Kakkar, L.R.: Talanta, *23*, 819 (1976)
11. Weiser, H.B.: J. Phys. Chem., *20*, 640 (1916)
12. Lelubre, R.: Ing. Chim. (Brussels), *25*, 101 (1941)
13. Newcomb, G., Markham, J.J.: Anal. Chim. Acta, *35*, 261 (1966)
14. Busev, A.I.: Analytical Chemistry of Molybdenum 1969 Ann Arbor, MI, USA, Ann Arbor-Humphrey Science
15. Elwell, W.T., Wood, D.F.: Analytical Chemistry of Molybdenum and Tungsten 1971 Oxford, Pergamon
16. McCay, L.W.: J. Amer. Chem. Soc., *56*, 2548 (1934)
17. Kato, H., Hosimiya, H.: J. Chem. Soc. Jpn., *60*, 1115 (1939); Chem. Abstr., *34*, 1585[7] (1940)
18. Rao, K.S., Vaidya, V.G.: Analyst (London), *100*, 512 (1975)
19. Liang, S.C., Hsu, H.H.P.: J. Chin. Chem. Soc., *17*, 90 (1950)
20. Haeringer, M., Goldstein, G., Lagrange, P., Schwing, J.P.: Bull. Soc. Chim. Fr., 723 (1967)
21. Yosida, Y.: J. Chem. Soc. Jpn., *61*, 130 (1940); Chem. Abstr., *34*, 4687[3] (1940)
22. Carrière, E., Dautheville, A.: Bull. Soc. Chim. Fr., 264 (1943)
23. Goto, H.: J. Chem. Soc. Jpn., *56*, 314 (1935); Chem. Abstr., *29*, 3936[9] (1935)
24. Přibl, R., Malát, M.: Coll. Czech. Chem. Commun., *15*, 120 (1950)
25. Spacu, P., Gheorghiu, C., Paralescu, I.: Acad. Repub. Pop. Rom. Stud. Cercet. Chim., *10*, 157 (1962)

26. Niericker, R., Treadwell, W.D.: Helv. Chim. Acta, *29*, 1472 (1946)
27. Borrel, M., Paris, R.: Anal. Chim. Acta, *4*, 267 (1950)
28. Knowles, H.B.: J. Res. Nat. Bur. Stand., *9*, 1 (1932)
29. Hoenes, H.J., Stone, K.G.: Talanta, *4*, 250 (1960)
30. Isibasi, S.: J. Chem. Soc. Jpn., *61*, 125 (1940); Chem. Abstr., *34*, 4687[1] (1940)
31. Das, M.: Diss. Abstr., *29B*, 1950 (1968)
32. Kharsan, R.S., Mishra, R.K.: Croat. Chem. Acta, *54*, 121 (1981)
33. Chandravanshi, B.S., Gupta, V.K.: Croat. Chem. Acta, *52*, 397 (1979)
34. Chaudhuri, N.K., Sarkar, A.K., Das, J.: Fresenius' Z. Anal. Chem. *254*, 365 (1971)
35. Eremin, Yu.G., Kolpikova, E.F., Rodionova, T.V.: Zh. Anal. Khim., *31*, 732 (1976)
36. Marchenko, P.V., Uzhvii, V.N.: Ukr. Khim. Zh., *31*, 612 (1965)
37. Eremin, Yu.G., Kolpikova, E.F., Dorfman, A.E.: Zh. Anal. Khim., *27*, 1297 (1972)
38. Chattopadhyay, S., Pal, B.K., Mitra, B.K.: Talanta, *22*, 431 (1975)
39. Collett, E., Eckardt, M.: Chem. Ztg., *33*, 968 (1910)
40. Trautman, W.: Chem. Ztg., *33*, 1106 (1910)
41. Gheorghiu, C., Radulescu-Grigor, E.: Rev. Chim. (Bucharest), *11*, 415 (1960)
42. Easton, A.J., Moss, A.A.: Miner. Mag., *35*, 995 (1966)
43. Rây, H.N.: Analyst (London), *78*, 217 (1953)
44. Koch, W., Brockmann, H.: Arch. Eisenhüttenwes., *34*, 441 (1963)
45. Golubsova, R.B., Shemyakin, F.M.: Zh. Anal. Khim., *4*, 232 (1949)
46. Rao, A.L.N., Tillu, M.: Indian J. Chem., *3*, 320 (1965)
47. Burriel-Marti, F., Vidan, A.M.: An. Real. Soc. Espan. Fis. Quim., Ser. B, *62*, 139 (1966)
48. Sajo, I., Förster, W., Rüdiger, H.A.: Neue Hütte, *12*, 500 (1967)
49. McNerney, W.N., Wagner, W.F.: Anal. Chem., *29*, 1177 (1957)
50. Naumann, H.C.: Metall (Berlin), *30*, 555 (1976)
51. Gardner, R.D., Ward, C.H., Ashley, W.H.: US At. Energ. Comm. Rept., LA 3058 (1964)
52. Cotiga, M.: Rev. Chim. (Bucharest), *19*, 615 (1968)

Chapter 5

Titrimetric Methods

Titrimetric methods serve as an alternate to gravimetric procedures and although perhaps not as accurate are more quickly performed, avoiding the necessity of drying a precipitate to constant weight. Provided a standard solution of titrating agent is available and a suitable means of end point detection found, many of the chemical reactions involving molybdenum can be applied towards a titration procedure. Most of the gravimetric methods discussed in the preceeding chapter have, for example, been adopted for titrations. Visual indicators, potentiometric readings and amperometric observations of the current flowing through polarized electrodes all serve as means of end point detection.

Molybdenum(VI) is the common, stable oxidation state for molybdenum. Reduction to molybdenum(V) or molybdenum(III) is easily achieved although Mo(III) once formed requires special handling as it readily reacts with atmospheric oxygen. Electrode potentials for molybdenum are:

$$\text{Mo(VI)} + 1\ e^- = \text{Mo(V)} \qquad E = +0.483 \text{ v vs. NHE* at 30 °C and } I = 0 \qquad [1]$$
$$\text{Mo(VI)} + 1\ e^- = \text{Mo(V)} \qquad E = +0.532 \text{ v vs. NHE* at 25 °C in 2 M HCl} \qquad [2]$$
$$\text{Mo(VI)} + 1\ e^- = \text{Mo(V)} \qquad E = +0.62 \text{ v vs. NHE* at 20 °C in 9 M HCl} \qquad [3]$$
$$\text{Mo(V)} + 2\ e^- = \text{Mo(III)} \qquad E = +0.30 \text{ v vs. NHE* at 20 °C in 9 M HCl} \qquad [3]$$
* NHE = normal hydrogen electrode

Because of complexation with anions present in acid solution, one expects potentials to differ in different acids and in different concentrations of the same acid. Small changes in potential also occur with changing temperature. Wünsch [3] measured molybdenum potentials in solutions varying from 7 M HCl to 15 M HCl and observed a steady increase in the Mo(V)/Mo(III) potential from +0.26 v vs. NHE in 7 M HCl to +0.56 v vs. NHE in 15 M HCl. The Mo(VI)/Mo(V) potential, however, remained essentially constant for acid strengths greater than 7 M HCl. It was +0.68 v vs. NHE in 7 M HCl and approximately +0.7 v vs. NHE in 15 M HCl [3].

Complexation reactions of molybdenum with ethylenediaminetetraacetic acid, EDTA, have analytical significance for both Mo(VI) and Mo(V). Acid base reactions of molybdate ion, as H_2MoO_4, are less suitable analytically and are subject to interference from any diverse ion that also possesses acid base properties. A constant and reproducible end point for the conversion of molybdate ion, MoO_4^{2-}, to paramolybdate ion, $Mo_7O_{24}^{6-}$, in the acid reaction

$$8\ H^+ + 7\ MoO_4{}^{2-} \rightarrow Mo_7O_{24}{}^{6-} + 4\ H_2O$$

of 8/7 = 1.143 for hydrogen to molybdenum is claimed provided the titration of molybdenum with standard HCl solution is carried out in the presence of a high salt concentration, 5 M HCl [4]. End point for this reaction is obtained by thermometric measurement and the procedure is applied to analysis of relatively pure molybdate salts.

1. Precipitation Titrations

1.1 With Various Metals

(i) Lead

Conditions for precipitation of molybdenum with lead are the same as those for gravimetric lead molybdate formation *viz.* a weak acid solution and absence of interfering ions. Acetic acid-ammonium acetate or hexamethylenetetraamine are commonly used as buffers at pH approximately 6. It is also possible to initially adjust the pH to 6 and proceed without addition of buffer. Various means of end point detection are possible for this titration. Generally, they center upon reactions involving excess lead ion present after the end point. Alizarin red is used as an adsorption indicator. It changes color when adsorbed upon the surface of freshly precipitated $PbMoO_4$ once excess lead is present in solution [5]. Other adsorption indicators work equally well [6, 7]. At best, the use of adsorption indicators requires skill in observing a color change even for an experienced analyst. Indicators which form colored lead complexes are preferred. Among these 4-(2-pyridylazo)resorcinol, PAR, is perhaps the best known [8]. In the titration of molybdenum with lead, the color change of PAR is from yellow-green to red at the end point [9]. Other complex forming indicators for lead in its titration of molybdate ion are listed in Table 5-1.

Concentration of lead during the course of titration can be followed potentiometrically with a Pb specific ion electrode [16]. Indirectly a fluoride sensitive electrode has been used for end point detection of Pb titrations. A small amount

Table 5-1. Indicators for the Titration of Molybdenum(VI) with Standard Lead(II) Titrant. Formation of a Colored Complex

Indicator	Condition	Color Change	Ref.
chlorophosphonazo III	pH 4–6	at end point→blue	[10]
1-(2,3-dihydroxy-4-pyridylazo)-4-benzene sulfonic acid	pH 4.5–7.0, pyridine	at end point orange→blue	[11]
diphenylcarbazone	neutral, 1:1 H_2O:EtOH, titrate with Mo(VI)	at end point→pink	[12]
dithizone	pH 4.8–5.5 HOAc-NH_4OAc	at end point green→red-violet	[13]
3-(2-lepidylazo)-naphthalene-1-sulfonic acid	pH 4.5–5.2 HOAc-NH_4OAc	(not available)	[14]
PAR	pH 6 HOAc-NH_4OAc	at end point yellow green→red	[8]
xylenol orange	pH 5–6 no buffer	at end point yellow→red	[15]

of chloride and fluoride ion is added to the molybdate solution before titration. After the end point, PbClF forms depleting the fluoride concentration. This is registered by the F^- ion electrode [17].

There are specific ion electrodes that respond directly to molybdate concentration [18]. The nitrate ion electrode also shows a maximum response to molybdate at pH 4.5 [19]. Titration with $Pb(NO_3)_2$, previously standardized against a molybdate solution of known concentration, is monitored at pH of 4.5 achieved by addition of either phosphoric acid or sodium hydroxide. Titrant must be added slowly, especially near the end point. Another specific ion electrode, although not used for Pb(II) titration, which responds directly to molybdenum is prepared from the ternary molybdenum complex with molybdenum(V), thiocyanate ion, and tetraethylammonium ion [20]. The unknown molybdenum sample must also be converted to the molybdenum(V) thiocyanate complex before a potential response occurs. Once this is done, the electrode responds predictably, i.e. the Nernst equation is obeyed, over the molybdenum concentration range from 10^{-2} to 5×10^{-8} M.

Titration of molybdenum with lead has been followed conductometrically, a break in the conductance vs. volume of lead titrant solution being observed at the end point [21]. Amperometric detection [22] and potentiometric detection with two polarizable electrodes [23] also signal the end point of lead molybdate titrations. Internal generation of lead, coulometric titration, for reaction with molybdenum and other ions is possible using a lead anode [24].

Anions in addition to molybdate which precipitate with lead must be absent from the sample. These include W(VI), Re(VI), Cr(VI), sulfate, thiosulfate, sulfide, fluoride, etc. Ions that complex with lead also interfere. These include tartrate, citrate, etc. Other ions also interfere under specific conditions, for example, nickel interferes if an amperometric end point is to be observed [25].

Following is the procedure of Cheng and Goydish for lead(II) titration of molybdenum(VI) [15].

The sample containing from 15 to 150 mg of molybdenum as MoO_4^{2-} and with no interfering ions is adjusted to pH 5–6 by addition of either ammonia, NH_3, or glacial acetic acid, $HC_2H_3O_2$. Add 5 to 6 drops of 0.1% (w/v) aqueous xylenol orange indicator. Titrate

Table 5-2. Metal Titrants for Molybdenum(VI)

Metal Titrant	Condition	End Point	Ref.
Ba^{2+}	pH 5.2, acetate	photometric, nitchromazo $\lambda = 650$ nm	[26]
Pb^{2+}	pH 6, acetate, 80 °C	visual, CuEDTA + 1-(2-pyridylazo)-2-naphthol, yellow→red	[9]
Hg_2^{2+}	1 M H_2SO_4, KSCN	amperometric, 2 Pt electrodes, +0.4 v	[27]
Ag^+	pH 6–7.2	potentiometric, Ag/SCE electrodes	[28]
Ag^+	neutral, 1 M KNO_3	amperometric, rotating Pt/SCE electrodes, +0.2 v	[29]
Tl^+	pH 5.8 0.1 M KNO_3 50% EtOH	potentiometric, Pt/Pb(Hg) electrodes	[30]
Mn(IV)	2.5 M H_2SO_4	visual ferroin, red→colorless	[31]

with standardized lead nitrate solution, $Pb(NO_3)_2$, of approximately 0.025 M concentration. Add titrant slowly and with stirring near the end point. Color change at the end point is from yellow-green to red. Millimoles of lead added equals mmoles of molybdenum reacted.

(ii) Other Metal Titrants

At one time or another most metal ions that form precipitates with molybdate ion were used for titration of molybdenum-containing solutions. Some of these are listed in Table 5-2. Note that each titrant ion is subject to interferences if other species are present in solution that also react, either through precipitate or complex formation.

1.2 With Organic Reagents

(i) 8-Hydroxyquinoline

The bromination reaction of 8-hydroxyquinoline (8-quinolinol, oxine) is quantitative and used to determine the amount of this material present in a sample. If the source of oxine is a metal oxine precipitate, then the procedure is applicable for determining the metal ion concentration also. Molybdenum oxine precipitate, described in chapter 4, is dissolved in acid, a known excess of BrO_3^- -Br^- added, and after bromination the excess bromine reacted with iodide. Liberated iodine is titrated with standard thiosulfate solution [32]. Chlorination rather than bromination of oxine is also possible [33]. If a known excess of oxine is used in the initial precipitation of molybdenum, then after filtering an aliquot of the filtrate is titrated with standard copper(II) nitrate solution at pH 8 using murexide indicator. Copper(II) reacts directly with the excess oxine and, after the end point, murexide gives a green colored copper(II) complex indicating quantitative reaction [34]. If the molybdenum oxine precipitate is dissolved in 2% (v/v) acetic anhydride-acetic acid solvent, it can be titrated as a diacidic base with 0.1 M $HClO_4$ in acetic acid as titrant. The end point is taken from the first potential break of the potentiometric titration curve followed with a glass/SCE electrode pair [35].

It is possible to titrate molybdenum(VI) directly with standardized 8-hydroxyquinoline at pH 3.7 to 5.5 using an amperometric end point [36]. It is best to titrate in a 25–30% (v/v) ethanol-water solvent, otherwise equilibrium is only slowly achieved after each addition of titrant. If the complexing agent cyclohexanediaminetetraacetic acid, DCTA, is present, interference from copper(II), iron(III), and tungsten(VI) is minimized when these ions are present in amounts approximately equal to the molybdenum concentration [37]. 8-Mercaptoquinoline (thioxine) is often used in place of 8-hydroxyquinoline for amperometric titration of molybdenum [38].

(ii) Other Organic Titrants

Table 5-3 lists several organic precipitating agents for molybdenum used in precipitation titrations.

Following is the procedure of Christova and Lihareva for precipitation of molybdenum(VI) with 8-hydroxyquinoline and amperometric end point [37].

Table 5-3. Organic Titrants for Molybdenum

Titrant	Condition	End Point	Ref.
benzyldithiocarbamate, K^+ salt	0.1 M HCl	amperometric, 2 Pt electrodes, +0.6 v	[39]
diantipyrylmethane	1 M H_2SO_4, SCN^-	amperometric, DME/SCE electrodes, -1.2 v	[40]
N-phenylbenzohydroxamic acid	1 M HCl	potentiometric, W/SCE electrodes polarized, 4 μa	[41]
tetramethyldiaminodiphenyl-methane	pH 3.6, acetate	amperometric, DME/SCE electrodes, -0.8 v	[42]
N-p-tolylthiobenzohydroxamic acid	5 M H_2SO_4	amperometric, DME/SCE electrodes, -0.5 v	[43]

The sample should contain between 0.2 and 1.2 mg Mo/mL. Accurately measure between 5 and 8 mL of sample into a titration cell. Add between 1 and 10 mL of 0.015 M cyclohexanediaminetetraacetic acid, $C_6H_{10}[N(CH_2CO_2H)_2]_2 \cdot H_2O$, depending upon the molybdenum concentration. The ratio of Mo:DCTA should be 1:1.5. Adjust the pH to 4.5 by dropwise addition of either 0.1 M hydrochloric acid, HCl, or 0.1 M ammonia, NH_3. Add 10 mL of acetic acid-ammonium acetate buffer solution (containing 10 mL glacial acetic acid, $HC_2H_3O_2$, and 8 g ammonium acetate, $NH_4C_2H_3O_2$, per 100 mL of solution). Add 10 mL 0.1 M ammonium fluoride, NH_4F. This serves as supporting electrolyte and masks aluminum ion if present in the sample. Deaerate the sample for 10 min by passing a stream of oxygen free nitrogen through the sample. Cover the sample and insert through openings in the cover a dropping mercury electrode, a saturated calomel electrode, and the tip of a 10 mL buret containing the standard oxine titrant. Set the potential of the DME at -0.8 v vs. SCE. (Here $E_{1/2}$ for Mo(VI) is -0.62 v vs. SCE.) Slowly add 0.0100 M oxine solution. (Prepare by dissolving 0.3629 g reagent grade oxine in 20 mL of 2 M $HC_2H_3O_2$. Dilute with distilled water to about 150 mL. Neutralize with 6 M NH_3 until the pH is 4.5 ± 0.1. Add additional distilled water until the final volume is exactly 250 mL.) Record the observed current on the amperometric unit after addition of each increment of titrant. Precipitate formation may not occur until well into the titration. Continue adding titrant until assured that the end point has been exceeded. Plot observed current, microamperes, vs., volume of added oxine, mL. Intersection of the extrapolated linear portions of the curve occurs at the end point volume. Two millimoles of oxine combine with each millimole of molybdenum. Calculate the amount of molybdenum present in the original sample.

2. Complexometric Titrations

Molybdenum(VI) reacts directly with ethylenediaminetetraacetic acid, EDTA, in solutions buffered at pH 4.5 [44,45] or in 50% (v/v) water-ethanol solutions at pH 5.2–5.6 [46]. Pyrocatechol-indigo carmine [44], diphenylcarbazone-methylene blue-vanadium(V) complex [46], or amperometric end point detection [45] all signal the end point of this titration. It is interesting that two molybdenum(VI) atoms attach to each EDTA molecule [47]. Generally, EDTA combines in a 1:1 ratio with metal ions.

More commonly used for molybdenum determination then the reaction of Mo(VI) with EDTA is the reaction of Mo(V) and EDTA. A known excess of ED-TA is added to a molybdenum(VI) solution followed by one or two grams of hydrazine sulfate, $H_2NNH_2 \cdot H_2SO_4$. Molybdenum is reduced by boiling the neutral or slightly acid solution for 5 to 10 min. The sample is cooled, adjusted to the appropriate pH for the metal ion titrant used in determining excess EDTA, and titrated with a standard metal ion solution. The ratio of molybdenum(V) to EDTA is also 2:1, EDTA combining with the Mo(V) dimer $Mo_2O_4^{2+}$ [48]. If iron(III) is used for the back titration, a sharper end point is observed potentiometrically if hydrazine chloride, $H_2NNH_2 \cdot 2HCl$, rather than hydrazine sulfate is used as the reducing agent [49]. Other metal titrants include copper(II) [50], zinc [51, 52], and zirconium(IV) [53]. Hydroxylamine hydrochloride, $NH_2OH \cdot HCl$, is not a satisfactory substitute for hydrazine in reducing molybdenum(VI) to molybden(V) if ED-TA is present. A ternary complex is formed, Mo(VI)-hydroxylamine-EDTA rather than molybdenum(V) EDTA [54]. Addition of tartrate ion prior to Mo(VI) reduction masks Ti, Nb, Ta, and W while addition of ammonium fluoride after reduction masks Th, Al, Ce, La, and U [55]. Silicon, if present, is also masked by fluoride [56]. It is possible to determine molybdenum with EDTA in binary mixtures containing zirconium-molybdenum, iron-molybdenum [57], or vanadium-molybdenum [58].

The following procedure of Přibil and Veselý is for titration of molybdenum (VI) with cyclohexanediaminetetraacetic acid, DCTA, a complexing agent similar to EDTA. In the presence of hydroxylamine hydrochloride and DCTA molybdenum(VI) forms a ternary complex sufficiently stable to allow excess DCTA to be titrated with standard zinc solution [59].

To a solution containing 5 to 10 mg Mo, add dropwise either 1.0 M hydrochloric acid, HCl, or 1.0 M ammonia, NH_3, until the solution is neutral, pH 7. Add 1 or 2 grams of hydroxylamine hydrochloride, $NH_2OH \cdot HCl$, and sufficient distilled water until the total volume is about 200 mL. Gently boil the solution for 5 to 10 min. Cool and adjust the pH to 4.5. Add 2 mL of acetic acid-ammonium acetate buffer solution (containing 10 mL glacial acetic acid, $HC_2H_3O_2$, and 8 g ammonium acetate, $NH_4C_2H_3O_2$, per 100 mL of solution). Add exactly 10.0 mL of 0.500 M DCTA, $C_6H_{10}[N(CH_2CO_2H)_2]_2 \cdot H_2O$, (Standardize DCTA with zinc.). Again gently boil the solution for 15 min. After cooling, add one gram hexamethylenetetraamine to raise the pH to 5.0 to 5.5. Add 2–3 drops 0.5% (w/v) xylenol orange dissolved in water and titrate the excess DCTA with 0.0500 M zinc solution. (Prepared by dissolving 0.3269 g pure zinc in a few mL of 6 M HCl. Dilute to approximately 50 mL with distilled water. Add 1 M NH_3 dropwise until the pH is 5. Add distilled water until a final volume of exactly 100 ml is achieved.) The end point is reached when the indicator changes to a red color. Up to a 50 fold excess of tungsten, not to exceed 300 mg W/ 500 mL solution, does not interfere with this procedure if a gross excess of hydroxylamine hydrochloride is used in the reduction step, 10 g rather than 1 g.

mmole Mo = total mmole – mmole Zn
           DCTA   for back titration.

Calculate the amount of molybdenum present in the sample.

3. Oxidation-Reduction Titrations

3.1 Reactions with Oxidizing Agents

(i) Prior Reduction of Molybdenum(VI)

For reaction with standard oxidizing titrants, Ce(IV), Mn(VII), Cr(VI), etc., molybdenum must first be reduced from its commonly found oxidation state of six to a lower oxidation state. Molybdenum(V) solutions are stable in air for several hours or longer and any of the common oxidizing titrants can quantitatively re-oxidize Mo(V) to Mo(VI). Mo(III) is air sensitive and to react quantitatively with oxidizing titrants, special precautions are necessary to exclude air. The molybdenum(IV) state although known to exist is rarely encountered in ordinary analytical procedures.

Various reducing agents, mostly metals and metal amalgams, successfully and quantitatively reduce molybdenum(VI) to either Mo(V) or Mo(III) [60–62]. It is sufficient to add a few mL of elemental mercury to a molybdenum(VI) solution in 3–6 M HCl and shake the mixture for a few minutes [63]. Other reducing agents, Sn(II), Ti(III), hydrazine, etc. also quantitatively reduce molybdenum. Frequently the reductant, be it metal or metal amalgam, is placed in a column and the molybdate solution passed through the column, the reduced form being collected at the column outlet. If Mo(V) is the desired reduction product, control of acid concentration and in some cases choice of acid is necessary. Bismuth, for example, produces Mo(V) quantitatively only if the hydrogen ion concentration, as HCl, is less than 1.7 M. Otherwise some Mo(III) is formed [64]. Likewise with silver reduction in 2 M HCl Mo(V) is formed but in 6 M HCl Mo(III) is the reduction product [65]. The stronger metallic reducing agents give Mo(III) exclusively as the reduction product. In these instances, however, it is desirable to exclude air, purge the system with an inert gas, otherwise the molybdenum(III) formed is immediately re-oxidized. Often molybdenum is determined indirectly by collecting the Mo(III) formed in a known amount of standard oxidant, for example, iron(III) solution. The iron(II) produced is then titrated with a standardized oxidizing agent, perhaps potassium dichromate, without fear of air oxidation. Table 5-4 lists various reducing agents for molybdenum(VI). These reductants are not unique for molybdenum. Other metal ions if present, e.g. W, U, V, Nb, Ta, Cr, etc., will also most likely be reduced and interfere with any subsequent quantitative procedure. Their prior removal is essential for quantitative determination of molybdenum in complex mixtures.

The following procedure of Stark uses mercury to reduce Mo(VI) to Mo(V) [66].

To the sample containing between 4 and 400 mg Mo in 250 mL of solution, add 25 mL of 12 M hydrochloric acid, HCl. Add 3–4 mL pure mercury. Stopper the flask and shake it vigorously for 10 min. The red-brown color of molybdenum(V) is observed. Green coloration, Mo(III), if present will be converted to Mo(V) once the flask is allowed to stand open in the air. Let the solid mercury(I) chloride formed settle and then pass the mixture through a filter containing a small piece of cotton to trap the solid. Wash the sides of the flask and the mercury(I) precipitate twice with 25 mL portions of 2.5 M HCl. Avoid as much as possible transferring any of the precipitate along with the reduced molybdenum solution. Combine the filtrate and washings and proceed with the titration of molybdenum(V).

Table 5-4. Reagents for the Prior Reduction of Molybdenum(VI) Before Titration with Standard Oxidizing Agents

Reductant	Condition	Oxidation State of Mo	Reference
Al(Hg)	0.25 M H_2SO_4	III	[67]
Bi	1.5 M HCl	V	[64]
Cd(Hg)	6 M HCl	III	[68]
Co	1 M H_2SO_4	III	[69]
hydrazine sulfate	1 M HCl, boil	V	[70]
Ag	2 M HCl, 75 °C	V	[71]
$NaBH_4$, 1 M NaOH	acidify with 1 M H_2SO_4	V	[72]
Zn(Hg)	4 M HCl	III	[73]

Stark also reduced Mo(VI) to Mo(III) using a zinc amalgam reductant [66].

To the sample containing between 1 and 100 mg Mo in 25 ml of solution, add 50 mL of 12 M hydrochloric acid, HCl. Add approximately 20 ml of zinc-mercury liquid amalgam. (Prepare by heathing on a water bath 15 g of zinc metal, 25 mL pure mercury, and 15 mL 6 HCl for one hour. Use a fume hood as mercury vapors are poisonous. Wash the mixture several times with 6 M HCl and separate any solid residue by placing the entire mixture in a separatory funnel. Draw off only the clean liquid mercury amalgam.) Stopper the flask and shake it vigorously for 5 min or until the green color indicating Mo(III) formation reaches its maximum intensity. Rapidly transfer the sample to a titration flask, wash the amalgam with 20 ml of 6 M HCl, and combine sample and wash solution. Titrate immediately with standard oxidizing titrant.

(ii) Titration of Reduced Molybdenum

All of the common oxidizing titrants quantitatively react with reduced molybdenum. Cerium(IV), for example, reacts with molybdenum(V) on a 1:1 mole for mole basis, each species gaining or losing one electron during the titration. With molybdenum(III) 3 moles of cerium(IV) react with each mole of molybdenum. Titrations are carried out in strong acid solution and, if Mo(III) is the form utilized, as rapidly as possible.

According to Stark, the following procedure is suitable for the titration of molybdenum(V) with Ce(IV) titrant [66].

Add to the reduced molybdenum solution of approximately 25–50 mL volume, 5–10 drops of 0.1% (w/v) N-phenylanthranilic acid (dissolved in 0.005 M NaOH). Titrate with 0.1 M standardized Ce(IV) sulfate in 1 M sulfuric acid, H_2SO_4. The indicator turns purple at the end point. One mmole of Ce(IV) is equivalent to one mmole of Mo(VI). Calculate the amount of molybdenum in the unknown sample.

Table 5-5 lists other indicators suitable for the cerium(IV)-molybdenum(V) titration.

Table 5-6 lists other titrants and conditions for reaction of standard oxidizing agents with reduced forms of molybdenum. End point detection is either po-

Table 5-5. Visual Indicators for Titration of Reduced Molybdenum with Standard Cerium(IV) Titrant

Indicator	Condition	Color Change	Ref.
diphenylamine-sulfonate	$H_2SO_4 + H_3PO_4$	at end point colorless→violet	[74]
ferroin	4 M $H_2SO_4 + H_3PO_4$	at end point red→pale blue	[75]
naphthidine-disulfonic acid	1 M $H_2SO_4 + H_3PO_4$	at end point colorless→pink	[76]
Nile blue	1 M H_2SO_4	at end point green→pink	[77]
N-phenyl-anthranilic acid	4 M HCl	at end point→purple	[66]
Ni phthalocyanine-sulfonic acid	1 M H_2SO_4	at end point blue→violet	[78]
promethazine HCl	0.5 M H_2SO_4	at end point colorless→pink	[79]

Table 5-6. Other Oxidizing Titrants for Titration of Reduced Molybdenum

Titrant	Condition	Indicator	Ref.
$KClO_3$	6 M HCl, Fe(III)	visual, methyl orange, red→colorless	[80]
chloramine T	8% HCl + 1.6% H_2SO_4, 75 °C	potentiometric, PT/SCE electrodes	[81]
$K_2Cr_2O_7$	2 M H_2SO_4, Fe(III)	visual, ferroin, red-orange→pale yellow	[82]
$K_2Cr_2O_7$	excess Fe(II) in 12 M H_3PO_4	potentiometric, 1st Fe(II) Fe(III), 2nd Mo(V)→Mo(VI)	[83]
$K_2Cr_2O_7$	2 M HCl + H_3PO_4	visual, diphenylamine,→violet	[84]
$FeCl_3$	1 M HCl	visual, methylene blue,→blue	[85]
$KMnO_4$	2 M H_2SO_4	visual, self-indicating,→purple	[86]
NH_4VO_3	1.5 M H_2SO_4	visual, chloropromazine, green→violet	[87]
$NaVO_3$	6 M $H_2SO_4 + H_3PO_3$	visual, Cu phthalocyaninesulfonic acid, blue→pink	[88]
NH_4VO_3	1 M HCl + $H_2C_2O_4$	visual, naphthidine, blue green→violet	[89]

tentiometric with a platinum indicating electrode or with an appropriate color-forming oxidation-reduction indicator.

3.2 Reactions with Reducing Agents

Some reducing agents can reduce molybdenum(VI) directly and quantitatively to a lower oxidation state. Of these chromium(II) is perhaps best known. Generally it is necessary to exclude air from a reaction of this type as oxygen readily attacks these titrants. A special titration apparatus is suggested for chromium(II) titrations [90] and end point detection is generally by potentiometric [91] or amperometric [92] means. Titration is at 70–85 °C. It is possible to titrate Mo(VI) with Cr(II) in the presence of W(VI) if oxalate is present in the solution [93]. Molybdenum(VI) is determined in the presence of vanadium(V) when titrated with Fe(II) [94]. Coulometric titrant generation avoids the problem of air oxidation. Titrants generated internally for reaction with molybdenum(VI) include chromi-

Table 5-7. Reducing Titrants for Titration of Molybdenum(VI)

Titrant	Condition	Indicator	Ref.
$CrCl_2$	8 M HCl+0.2 M $H_2C_2O_4$, 85 °C	potentiometric, Pt/SCE electrodes	[91]
ferrocene	50% 12 M HCl-acetone	amperometric, Pt/alkaline permanganate electrodes, 0 v	[100]
$FeSO_4 \cdot (NH_4)_2SO_4$	12 M H_3PO_4	visual, cacotheline, add near end point, yellow→pink	[101]
Nb^{3+} in 9 M H_2SO_4	6 M HCl, CO_2 purge, T<20 °C	visual, phenosafranine →colorless	[102]
$SnCl_2$ in 8 M HCl	8 M HCl	visual, ferroin+safranine T, blue→gray→yellow	[103]
VCl_2	4 M HCl+H_3PO_4	amperometric, DME/SCE electrodes, −0.7 v	[104]
$V_2(SO_4)_3$	6 M HCl	visual, phenosafranine→violet	[105]

Table 5-8. Titrimetric Determination of Molybdenum in Specific Materials

Material	Titrant	Reference
molybdenum ore		
molybdenite	Hg_2Cl_2	[27]
powellite	$BaCl_2$	[26]
ore concentrate	N-phenylbenzohydroxamic acid	[106]
ferrous alloys		
ferromolybdenum	$Pb(NO_3)_2$	[107]
ferromolybdenum	$KMnO_4$	[108]
ferromolybdenum	EDTA	[109, 110]
ferromolybdenum	N-phenylbenzohydroxamic acid	[111]
Cr-Mo steel	$CrSO_4$	[112]
steel	VCl_2	[104]
steel	ferrocene	[113]
steel	benzyldithiocarbamate, K^+ salt	[39]
steel	8-mercaptoquinoline	[114]
steel	N-phenylbenzohydroxamic acid	[41]
nonferrous alloys		
niobium alloy	EDTA	[115]
nickel alloy	Ce(IV)	[116]
nickel alloy	$K_2Cr_2O_7$	[117]
tungsten alloy	EDTA	[55, 118]
tungsten alloy	ferrocene	[119]
miscellaneous samples		
molybdenum carbide, nitride, and silicide	citric acid	[120]
zirconium boride	EDTA	[121]
molybdenum containing catalyst (Al-Co-Mo)	EDTA	[122]
molybdenum containing frits (alk. earth oxides, SiO_2)	N-phenylbenzohydroxamic acid	[123]
molybdenum containing glaze	$SnCl_2$	[103]

um(II) [95], iron(II) [96], tin(II) [97], and titanium(III) [98, 99]. Table 5-7 lists some titrants used in the reduction of molybdenum(VI).

4. Applications to Specific Materials

Nearly every molybdenum containing substance has, provided sufficient molybdenum is present, been determined by a titrimetric procedure. Titrations are rapid, accurate, and precise. Provided interfering ions can be removed or masked, they yield satisfactory results in most instances. Table 5-8 lists several applications of titrimetric procedures for molybdenum in various materials.

References

1. El-Shamy, H.K., El-Aggan, A.M.: J. Amer. Chem. Soc., 75, 1187 (1953)
2. Speranskaya, E.F., Mertsalova, V.E., Kulev, I.I.: Usp. Khim., 35, 2129 (1966)
3. Wünsch, G.: Talanta, 27, 649 (1980)
4. Kiba, N., Takeuchi, T.: Anal. Chim. Acta, 66, 75 (1973)
5. Raĭkhinshtein, Tz., Korobov, N.: J. Gen. Chem. USSR, 2, 661 (1932)
6. Cândea, C., Murgulescu, I.G.: Ann. Chim. Anal. Chim. Appl., 18, 33 (1936)
7. Henkel, H.: Fresenius' Z. Anal. Chem., 119, 326 (1940)
8. Lassner, E., Scharf, R., Püschel, R.: Fresenius' Z. Anal. Chem., 165, 29 (1959)
9. Püschel, R., Lassner, E., Scharf, R.: Fresenius' Z. Anal. Chem., 163, 104 (1958)
10. Avgushevich, I.V., Chernyshova, T.V., Kulikova, E.S., Luken, A.M.: Khim. Tverd. Topl., 56 (1976); Chem. Abstr., 86, 57815z (1977)
11. Gupta, J.P., Garg, B.S., Singh, R.P.: J. Indian Chem. Soc., 54, 1100 (1977)
12. Deshmukh, G.S.: Bull. Chem. Soc. Jpn., 29, 27 (1956)
13. Ueda, S., Yamamoto, Y., Takenouchi, H.: Nippon Kagaku Zasshi, 88, 1299 (1967)
14. Chadha, R.C., Garg, B.S., Lata, S., Singh, R.P.: J. Indian Inst. Sci., 62 B, 173 (1980); Anal. Abstr., 42, 1B14 (1982)
15. Cheng, K.L., Goydish, B.L.: Microchem. J., 13, 35 (1968)
16. Chao, E.E., Cheng, K.L.: Talanta, 24, 247 (1977)
17. Efstathion, C.E., Haljiioannou, T.P.: Anal. Chim. Acta, 109, 319 (1979)
18. Malik, W.U., Srivastava, S.K., Bansal, A.: Anal. Chem., 54, 1399 (1982)
19. Bernal, J.J., Pardo, R., Barrado, E.: Anal. Lett., 13A, 241 (1980)
20. Fogg, A.G., Kumar, J.L., Burns, D.: Anal. Lett., 7, 629 (1974)
21. Rother, E., Jander, G.: Z. Angew. Chem., 43, 930 (1930)
22. Aylward, G.H.: Anal. Chim. Acta, 14, 386 (1956)
23. Rusina, O.N., Gorbatkova, B.Kh.: Ukr. Khim. Zh., 43, 982 (1977)
24. Tutundzic, P.S., Stojkovic, D.J.: Zh. Anal. Khim., 21, 436 (1966)
25. Babko, A.K., Volkova, A.I.: Zavod. Lab., 24, 135 (1958)
26. Kreshkov, A.P., Kuznetsov, V.V., Mezhlumyan, P.G.: Zh. Anal. Khim., 29, 1349 (1974)
27. Tarayan, V.M., Acharyan, G.S., Shaposhnikova, G.N.: Arm. Khim. Zh., 27, 279 (1974)
28. Shivahare, G.C.: Fresenius' Z. Anal. Chem., 219, 187 (1966)
29. Songena, O.A., Zakharov, V.A., Mambetkaziev, E.A.: Zh. Anal. Khim., 23, 453 (1968)
30. Yoshimura, C., Uno, S.: Bunseki Kagaku, 12, 42 (1963)

31. Mandal, S.K., Sant, B.R.: Indian J. Chem., *19A*, 389 (1980)
32. Montequi, R., Doadrio, A.: An. Fis. Quim. (Madrid), *1*, 311 (1947)
33. Spacu, P., Gheorghiu, C., Paralescu, I.: Fresenius' Z. Anal. Chem., *195*, 321 (1963)
34. Ivanov, N., Karadakov, B.P.: God. Khim. Tekhnol. Inst., *11*, 1 (1964); Chem. Abstr., *65*, 12855c (1966)
35. Kreshkov, A.P., Kuznetsov, V.V., Bakalova, L.M., Yakovlev, P.Ya.: Zavod. Lab., *39*, 1056 (1973)
36. Manok, F., Kovacs, C.: Stud. Univ. Babes-Bolyai, Ser. Chem., *9*, 85 (1964); Chem. Abstr., *61*, 12623g (1964)
37. Christova, R., Lihareva, N.: Fresenius' Z. Anal. Chem., *259*, 348 (1972)
38. Suprunovich, V.I., Usatenko, Yu.I.: Zh. Anal. Khim., *20*, 800 (1965)
39. Galushko, S.V., Usatenko, Yu.I.: Zavod. Lab., *40*, 776 (1974)
40. Zhivopistsev, V.P., Sadakov, G.A.: Uch. Zap. Perm. Gos. Univ., No. 141, 174 (1966); Chem. Abstr., *68*, 56321b (1968)
41. Boeva, L.V., Kimstach, V.A., Bagdasarov, K.N.: Zh. Anal. Khim., *35*, 313 (1980)
42. Papfil, M., Furnica, M.: An. Stiint. Univ. Al. I. Cuza Iasi, Sect. Ic., *10*, 17 (1964); Chem. Abstr., *63*, 15547b (1965)
43. Gallai, Z.A., Sheina, N.M., Polikarpova, N.V., Vilkova, O.M.: Zh. Anal. Khim., *31*, 921 (1976)
44. Busev, A.I., Chang, F.: Vestn. Mosk. Univ., Ser. Mat. Mekh. Astron. Fiz. Khim., *14*, 203 (1959); Chem. Abstr., *54*, 8457d (1960)
45. Zhdanov, A.K., Umurzakova, I.A.: Nek. Vopr. Khim. Tekhnol. Fiz. Khim. Anal. Neorgan. Sistem, Akad. Nauk Uz. SSR, Otd. Khim. Nauk, *163* (1963); Chem. Abstr., *61*, 1259e (1964)
46. Sajo, I.: Fresenius' Z. Anal. Chem., *199*, 16 (1963)
47. Kula, R.J.: Anal. Chem., *38*, 1581 (1966)
48. Mitchell, P.C.H.: Quart. Rev., *20*, 103 (1966)
49. Khristova, R., Nonova, D.: Acta Chim. Acad. Sci. Hung., *81*, 433 (1974)
50. Lassner, E., Scharf, R.: Fresenius' Z. Anal. Chem., *167*, 114 (1959)
51. Busev, A.I., Chang, F.: Zh. Anal. Khim., *14*, 445 (1959)
52. Headridge, J.B.: Analyst (London), *85*, 379 (1960)
53. Klygin, A.E., Kolyada, N.S., Zavrazhnova, D.M.: Zh. Anal. Khim., *16*, 442 (1961)
54. Lassner, E., Schedle, H.: Talanta, *15*, 623 (1968)
55. Lassner, E., Scharf, R.: Fresenius' Z. Anal. Chem., *168*, 429 (1959)
56. T'ao, T.C., Yank, W.: Hua Hsuch Tung Pao, (5), 57 (1966); Chem. Abstr., *65*, 19302e (1966)
57. Akent'eva, N.A., Zhdanov, A.K., Zakirov, B.G.: Uzb. Khim. Zh., *14*, 15 (1970)
58. Zhdanov, A.K., Barkhudar'yan, A.A.: Uzb. Khim. Zh., (6), 6 (1976)
59. Přibil, R., Veselý, V.: Talanta, *17*, 170 (1970)
60. Höltje, R., Geyer, R.: Z. Anorg. Allg. Chem., *246*, 243 (1941)
61. Yoshimura, T.: J. Chem. Soc. Jpn., *73*, 122 (1952)
62. Becker, J., Coetzee, C.J.: Analyst (London), *92*, 166 (1967)
63. Furman, N.H., Murry, W.M.: J. Amer. Chem. Soc., *58*, 1689 (1936)
64. Ankudimova, E.V.: Tr. Kom. Anal. Khim., Akad. Nauk SSSR, Otd. Khim. Nauk, *5*, 197 (1954); Chem. Abstr., *49*, 12187d (1955)
65. Hiskey, C.F., Springer, V.F., Meloche, V.W.: J. Amer. Chem. Soc., *61*, 3125 (1939)
66. Stark, J.G.: J. Chem. Educ., *46*, 505 (1969)
67. Bhaskare, C.K., Surekha, D., Ganage, K.N.: J. Indian Chem. Soc., *55*, 177 (1978)
68. Speranskaya, E.F., Mertsalova, V.E.: Izv. Vyssh. Uchebn. Zaved., Khim. Khim. Tekhnol., *8*, 893 (1965); Chem. Abstr., *64*, 13742g (1966)
69. Polotebnova, N.A., Branover, L.M.: Uch. Zap. Kishinev. Gos. Univ., *14*, 119 (1954); Chem. Abstr., *52*, 5203h (1958)

70. Rao, G.G., Suryanarayana, M.: Fresenius' Z. Anal. Chem., *168*, 177 (1959)
71. Grubitsch, H., Halvorsen, K., Schindler, G.: Fresenius' Z. Anal. Chem., *173*, 414 (1960)
72. Andreev, F.I., Khain, V.A., Zakharov, M.S.: Zavod. Lab., *45*, 292 (1979)
73. Lundell, G.E.F., Knowles, H.B.: Ind. Eng. Chem., *16*, 723 (1924)
74. Yoshimura, C.: Nippon Kagaku Zasshi, *76*, 883 (1955)
75. Dikshitulu, L.S.A., Rao, G.G.: Fresenius' Z. Anal. Chem., *202*, 344 (1964)
76. Gowda, H.S., Shakunthala, R.: Anal. Chim. Acta, *91*, 399 (1977)
77. Rao, N.V., Avadhanulu, A.B., Sarma, C.B.R., Khrishna, K.S.: J. Indian Chem. Soc., *52*, 1048 (1975)
78. Rao, G.G., Rao, N.V.: Fresenius' Z. Anal. Chem., *190*, 213 (1962)
79. Gowda, H.S., Ahmed, S.A.: J. Indian Chem. Soc., *58*, 561 (1981)
80. Rao, G.G., Radhakrishnamurty, C.: Bull. Chem. Soc. Jpn., *45*, 3437 (1972)
81. Doležal, J., Moldán, B., Zýka, J.: Collect. Czech. Chem. Commun., *24*, 3769 (1959)
82. Sriramam, K.: Talanta, *19*, 1445 (1972)
83. Muralikrishna, U., Rao, G.G.: Talanta, *15*, 143 (1968)
84. Weiner, R., Boriss, P.: Fresenius' Z. Anal. Chem., *160*, 343 (1958)
85. Sagi, S.R., Rao, G.G.: Acta Chim. Acad. Sci. Hung., *38*, 89 (1963)
86. Speranskaya, E.F., Mertsalova, V.E.: Khim. Khim. Tekhnol. (Alma-Ata), *2*, 89 (1964); Chem. Abstr., *64*, 1344d (1966)
87. Gowda, H.S., Shakunthala, R.: Talanta, *13*, 1375 (1966)
88. Rao, G.G., Sastri, T.P.: Fresenius' Z. Anal. Chem., *167*, 1 (1959)
89. Gowda, H.S., Shakunthala, R.: Anal. Chim. Acta, *97*, 385 (1978)
90. Flatt, R., Sommer, F.: Helv. Chim. Acta, *25*, 684 (1942)
91. El-Shamy, H.K., Barakat, M.F.: Egypt. J. Chem., *2*, 191 (1959)
92. Nikolaeva, E.R., Agasyan, P.K., Tarenova, K.Kh., Kadykova, T.S.: Zh. Anal. Khim., *25*, 119 (1970)
93. Goryushina, V.G., Cherkashina, T.V.: Zavod. Leb., *14*, 255 (1948)
94. Rao, G.G., Dikshitulu, L.S.A.: Talanta, *10*, 1023 (1963)
95. Kostromin, A.I., Mosolov, V.V., Makarova, L.L.: Zh. Anal. Khim., *27*, 2114 (1972)
96. Agasyan, P.K., Tarenova, K.Kh., Nikolaeva, E.R., Katina, R.M.: Zavod. Lab., *33*, 547 (1967)
97. Agasyan, P.K., Nikolaeva, E.R., Tarenova, K.Kh.: Zavod. Lab., *35*, 1034 (1969)
98. Yen, H.Y., Liu, Y.H.: Hua Hsueh Tung Pao, *17*, 279 (1966); Chem. Abstr., *66*, 25801n (1967)
99. Nikolaeva, E.R., Agasyan, P.K., Tarenova, K.Kh., Boikova, S.I.: Vestn. Mosk. Univ., Ser. II, *23*, 73 (1968); Chem. Abstr., *70*, 43771q (1969)
100. Silomatin, V.T., Yakovlev, P.Ya., Artemova, T.N., Lapshina, L.A.: Zh. Anal. Khim., *28*, 2197 (1973)
101. Krishnamurthy, N., Pullarao, Y.: Indian J. Chem., *14A*, 1023 (1976)
102. Sen, B.K., Maity, R.K., Gupta, R.N., Bandyapadhyay, P.: Anal. Chim. Acta, *81*, 173 (1976)
103. Henze, G., Geyer, R., Lahl, W.: Neue Hütte *14*, 54 (1969)
104. Gusev, S.I., Nikolaeva, E.M.: Zh. Anal. Khim., *19*, 715 (1964)
105. Mandal, S.K.: Talanta, *26*, 133 (1979)
106. Borovaya, N.S., Kudelina, N.E., Tserkovnitskaya, I.A.: Vestn. Leningr. Univ. Fiz. Khim., (2), 138 (1978), Chem. Abstr., *89*, 139857r (1978)
107. Lukianets, I.G., Kulish, N.G.: Izv. Vyssh. Uchebn. Zaved. Khim. Khim. Tekhnol., *24*, 291 (1981); Anal. Abstr., *41*, 5B163 (1981)
108. Mabee, H.C.: Can. Chem. J., *2*, 132 (1918)
109. Endo, Y., Tomori, T.: Bunseki Kagaku, *11*, 1310 (1963)
110. Nikitina, E.I., Aduanova, N.N.: Zavod. Lab., *31*(6), 654 (1965)

111. Boeva, L.V., Bagdasarov, K.N., Kimstock, V.A.: Zavod. Lab., *42*, 515 (1976)
112. Brintzinger, H., Rost, B.: Fresenius' Z. Anal. Chem., *115*, 250 (1939)
113. Yakovlev, P.Ya., Solomatin, V.T., Bakalova, L.M.: Zavod. Lab., *40*, 1046 (1974)
114. Bogovina, V.I., Novak, V.P., Mal'tsev, V.F.: Zh. Anal. Khim., *20*(9), 951 (1965)
115. Polyak, L.Ya., Bashkirova, I.S.: Zh. Anal. Khim., *21*, 682 (1966)
116. Montgomery, E.L.: Anal. Chim. Acta, *121*, 85 (1980)
117. Hsuan, W.-K., Peng, L.-Y.: Fen Hsi Hsu Hsueh, *8*(4), 332 (1980); Anal. Abstr., *42*, 2B148 (1982)
118. Popova, O.I., Seroya, O.G.: Zh. Anal. Khim., *23*(5), 791 (1968)
119. Yakovlev, P.Ya., Solomatin, V.T., Bakalova, L.M.: Zavod. Lab. *40*, 915 (1974)
120. Kugai, I.N., Nazarchuk, T.N.: Zh. Anal. Khim., *17*, 1082 (1962)
121. Nazarchuk, T.N., Kugai, L.N., Galadzhii, O.F.: Zh. Anal. Khim., *22*(2), 240 (1967)
122. Uvarova, E.I., Rik, V.M.: Khim. Tekhnol. Topliv. Masel, *9*(5), 67 (1964); Chem. Abstr., *61*, 4964d (1964)
123. Sokolova, A.A., Mukhina, Z.S.: Zavod. Lab., *34*, 1060 (1968)

Chapter 6

Colorimetric Methods

Colorimetry, the application of absorbance measurements upon visible, colored species to determine the amount of component present, is a well established chemical discipline. Absorbance values measured with standard solutions containing known amounts of molybdenum are compared with similar measurements obtained from unknown solutions and the molybdenum content of the unknown determined through the relation between absorbance and concentration. Molybdenum has the ability to form colored complexes with many reagents and choices of procedure for a particular determination are numerous. Ideally the color forming reagent is both specific for the metal of interest and produces a highly colored complex. No one reagent meets these goals for molybdenum. Thiocyanate ion, reacting with molybdenum(V) and toluene-3,4-dithiol, reacting with molybdenum(VI), are the most widely used reagents. Many other reagents are available.

Other procedures depending upon measurements within the visible region of the electromagnetic spectrum, fluorimetry, polarimetry, etc., although they have been employed for determining molybdenum, are not as widely used in analysis of molybdenum containing samples and are not discussed in this report. Similarly, procedures employing ultraviolet and infrared measurements are omitted.

1. Color Forming Reagents

1.1 Thiocyanate Ion

Formation of the orange-red thiocyanate complex of molybdenum(V) is the most common colorimetric procedure for determining molybdenum. A reducing agent is added to an acidified solution of molybdenum(VI) in the presence of thiocyanate ion and the molybdenum(V) thiocyanate complex extracted into an organic solvent. Absorbance measurements are made at about 470 nm, exact wavelength varying somewhat with varying conditions. Molar absorptivity of molybdenum(V) thiocyanate complex is approximately 15,000 L mol^{-1} cm^{-1}, again, varying depending upon conditions and choice of extracting agent. Color is sufficiently stable in the organic phase for measurements to be made upon a standard series of known and the unknown molybdenum solutions. In spite of its widespread use, there is a lack of agreement regarding the composition of molybdenum(V) thiocyanate complex [1]. The formula $MoO(NCS)_3$ fits the stoichiometry of an oxygenated molybdenum(V) compound. The possible number of thiocyanate groups, however, is variable [2] with from one to five thiocyanate groups adding to the molybdenum cation in a stepwise fashion. The species of analytical interest probably contains four or five bound thiocyanate groups. The complex is then negatively charged thus allowing for formation of ion association complexes, for example, $[R_4N]^+$ $[MoO(NCS)_4]^-$ [3]. It is possible that the complex is a dimer, perhaps $[(MoO_2)_2(NCS)_6]^{4-}$ [4] or $[(MoO_2)_2NCS]^+$ [5]. Some claim molybdenum present in the complex is Mo(VI) rather than Mo(V) or a mixture of Mo(V) and Mo(VI) depending upon the choice of reducing agent used in the procedure [6]. Still others find no evidence for molybdenum oxygen bonds preferring the formula $[Mo(NCS)_6]^-$ [7]. For the analyst it is not necessary that exact composition be known as long as a linear relation holds between molybdenum concentration and absorbance. Still, the question is an interesting one.

Much work has gone into selecting a proper reducing agent for converting molybdenum from its plus six to its plus five oxidation state. Tin(II) in strong acid solution, 9–12 M hydrochloric acid, is commonly used although further reduction to molybdenum(III) by Sn(II) occurs [8,9]. The presence of perchloric acid with 3 M HCl and Sn(II) is said to achieve reduction to only the +5 state for molybdenum [3]. Addition of Sn(II) dissolved in diethyleneglycol rather than hydrochloric acid is suggested [10] as is the addition of hydroquinone to prevent air oxidation of Mo(V) once it is formed [11]. Titanium(III) is substituted for Sn(II) but it, too, causes some reduction to Mo(III) along with Mo(V) [12]. Several milder reducing agents have been suggested with the anticipation of forming only Mo(V) and not lower molybdenum states. Among these ascorbic acids is the most common [13]. Thiourea reduces molybdenum to Mo(V) [14] but, in addition, also complexes with molybdenum thus competing with thiocyanate [15]. Other reducing agents include iodide ion [16], hydrazine [17], and ferrocene [18]. The latter reagent, ferrocene, is fairly specific for molybdenum reduction only.

Many claim that these mild reducing agents do not quantitatively convert Mo(VI) to Mo(V). One finds procedures incorporating mixtures of reducing agents. In addition, iron(III) or Cu(II) salts are frequently added along with a reducing agent to improve the reduction processes. The mechanism for this improvement is not fully explained. With iodide ion as reducing agent, sodium sulfite is added to consume iodine formed as a result of reaction with molybdenum(VI).

Ions similar to molybdenum that also complex with thiocyanate interfere with this procedure. Most notable among these are rhenium, vanadium, tungsten, and uranium. Other ions that exhibit colored thiocyanate complexes, iron(III), cobalt, nickel, etc., or in some other way interact with thiocyanate are best removed or masked especially if they are present in excess relative to the amount of molybdenum in the sample [19]. It is possible to determine both molybdenum and tungsten in the same solutions by measuring at both wavelengths of maximum absorbance, 470 nm and 403 nm respectively. One then solves the necessary simultaneous equations [20]. If a mild reducing agent, ascorbic acid or thiourea, is used for molybdenum(V) formation, attack upon tungsten is not as readily achieved and its interference lessened [21]. Interference from aluminum is also lessened when a mild reducing agent for molybdenum is employed [14]. Ferrocene,di-π-cyclopentadienyl-iron(II), is reported to be a highly specific reducing agent for molybdenum and is used in determining molybdenum by the thiocyanate procedure in the presence of niobium, rhenium, and tungsten [18]. Interference from cobalt, nickel, rhenium, titanium, tungsten, and vanadium is removed if molybdenum(V) thiocyanate complex formed by hydrazine reduction is extracted with 1% (w/v) tribenzylamine in chloroform and then the organic layer back extracted with 1% (w/v) Sn(II) in 1 M HCl. Molybdenum remains in the $CHCl_3$ layer [7]. Tártrate, citrate, oxalate, fluoride, or EDTA are frequently added as masking agents prior to thiocyanate formation [18, 22].

A procedure for molybdenum determination as the thiocyanate complex is available using continuous flow injection technique [23].

The procedure of Solomatin, Yakovlev, Lapshina, and Artemova [18] uses ferrocene to selectively reduce molybdenum in the presence of ions which generally interfere with molybdenum thiocyanate measurement. Using this procedure, much higher concentrations of Nb, rare earths, Re, Ti, W, Zn, and Zr can be tolerated than are possible using ascorbic acid. Reduction with ferrocene is optimum in a solution of 0.25 M HCl and 10–14 M acetic acid. Molar absorptivity for molybdenum by this procedure is 14,500 L mol^{-1} cm^{-1} at 465 nm, the wavelength of maximum absorbance. Beer's law, indicating a linear relationship between absorbance and concentration, is followed for solutions containing from 0.2 to 4.0 µg Mo/mL. Details of the procedure follow.

A solution of *known molybdenum concentration* is prepared by dissolving exactly 0.100 g pure Mo metal in 10 mL of 8 M nitric acid, HNO_3. After the molybdenum is in solution, boil the sample to expel oxides of nitrogen. Carefully evaporate the solution to dryness and dissolve the residue with 12 M hydrochloric acid, HCl. Dilute the sample to exactly 100 mL with 12 M HCl. Exactly 10.00 mL of this solution is diluted to 100 mL in a volumetric flask with 6.0 M HCl. This diluted stock solution contains 0.100 mg Mo/mL of solution and an acid concentration of 6.6 M HCl.

To prepare a calibration plot, place between 0.100 and 1.50 mL of diluted stock solution into separate 50 mL volumetric flasks. This results in a series of standards containing between 10.0 and 150 µg Mo/50 mL. To the 10.0 µg Mo/50 mL solution add 1.95 ml 6.0 M HCl. To the 150 µg Mo/50 mL solution add 0.43 mL 6.0 M HCl. Samples containing intermediate amounts of molybdenum should have the appropriate intermediate amount of 6.0 M HCl added. Upon dilution to 50 mL, all flasks will be 0.25 M in HCl. Add to each flask 10 mL of glacial acetic acid, $HC_2H_3O_2$. Mix the contents of each flask and add to each 0.40 mL ferrocene solution (0.01 M ferrocene in ethanol). Again mix the contents of each flask and allow them to stand for five minutes. Now add to each flask 4.0 mL of 20% (w/v) potassium thiocyanate, KSCN, solution. Dilute each flask to a final volume of 50 mL. Mix the contents of each flask and allow them to stand for 20 min. At the end of that time, record the absorbance of each standard against a reagent blank prepared in the manner described but containing no molybdenum. The color is stable for up to five hours. Maximum absorbance is measured at 465 nm.

An *unknown sample* following appropriate dissolving is acidified with HCl, if not already done, and a specific aliquot taken which provides for a molybdenum concentration somewhere within the range of the calibration curve. Place the sample aliquot in a 50 mL volumetric flask. Add sufficient additional 6.0 M HCl to adjust the final HCl concentration to 0.25 M and proceed with addition of other reagents as outlined for the molybdenum standards. Concentration of molybdenum in the aliquot is read, following absorbance measurement, from a calibration curve of absorbance vs. amount of molybdenum in the standards. Following correction for dilution, the amount of molybdenum in the original samples is reported.

Current emphasis on molybdenum thiocyanate complexes centers about formation of ternary complexes. For the most part, these are ion association complex-

Table 6-1. Auxiliary Ligands for Ternary Complex Formation with Molybdenum(V) Thiocyanate Complex

Reagent	Condition	Wavelength of Maximum Absorbance, nm / Molar Absorptivity, $L\ mol^{-1}\ cm^{-1}$ / Specific Absorptivity,[a] $mL\ g^{-1}\ cm^{-1}$	Mole Ratio Mo:SCN:Rgt	Ref.
2-benzylaminopyridine, 0.2 M CHCl$_3$	0.7 M H$_2$SO$_4$, ascorbic acid, heat, extract CHCl$_3$	465 2.0×10^4 0.21	1:3:2	[24]
chlorpromazine HCl, 0.1% (w/v) aqueous	2.0 M HCl, ascorbic acid, 20 min, extract CHCl$_3$	465 1.6×10^4 0.17	1:4:1	[25]
crystal violet, 2×10^4 M toluene	1.8 M H$_2$SO$_4$, ascorbic acid, 20 min, extract toluene	595 2.3×10^5 2.40	1:5:2	[26]
N,N'-diphenyl-p-toluamidine HCl, 1% (w/v) aqueous	4 M HCl, ascorbic acid, extract benzene	470 1.8×10^4 0.18	1:2:2	[27]
ethyl xanthate, 1% (w/v) aqueous	7 M HCl, extract acetophenone	470 1.4×10^4 0.15	1:4:2	[28]
lobeline HCl, 0.1 M aqueous	1 M HCl, hydrazine, boil, extract CHCl$_3$	465 1.4×10^4 0.14	1:5:2	[29]
nitron sulfate, 0.05% (w/v) aqueous	3 M HCl, ascorbic acid, Fe(II), extract CHCl$_3$	465 1.5×10^4 0.16	2:10:3	[30]
pecazine, 0.02 M CHCl$_3$	1.3 M HCl, ascorbic acid, 15 min, extract CHCl$_3$	465 1.8×10^4 0.18	1:4:1	[31]
tetraphenylarsonium chloride, 0.025 M aqueous	2 M H$_2$SO$_4$, ascorbic acid, Ti(III), 20 min, extract CHCl$_3$	470 1.7×10^4 0.18	1:4:1	[32]
rhodamine 6 G, 0.01% (w/v) aqueous	0.6 M HCl, 60 min, 1% gelatin	570 2.6×10^5 2.71	1:5:2	[33]

[a] Specific absorptivity corresponds to absorbance that would be observed from a solution containing 1 μg Mo/mL when measured in a cell of one centimeter path length

es in which the negatively charged molybdenum thiocyanate is joined to a cationic species, frequently a protonated, substituted amine or similar ion. Not only is the absorptivity of the overall complex enhanced, allowing for a more sensitive molybdenum procedure, but the ternary is extractable into chloroform, allowing for improved selectivity. Molybdenum thiocyanate complex alone is not soluble in chloroform and is generally extracted with 2-pentanol or methyl isobutyl ketone. Table 6-1 lists some of the auxiliary reagents associated with molybdenum thiocyanate.

The procedure of Fogg, Kumar, and Burns uses tetraphenylarsonium ion to form a ternary complex with molybdenum and thiocyanate [32]. The complex is extracted into chloroform where its absorbance is measured at 470 nm. Molar absorptivity is 17,400 L mol^{-1} cm^{-1} and the complex is stable for 24 h. Absorbance is directly proportional to concentration, Beer's law, for 0.8 to 6.0 μg Mo/mL. Reduction of molybdenum is achieved with ascorbic acid and titanium(III) chloride. Extraction is with chloroform containing a small amount of hydroquinone to prevent unwanted oxidation products in the solvent. Iron(III) is added to each molybdenum standard and sample if not already present. Addition of oxalic acid prevents interference from tungsten (up to 40 fold excess over Mo) and niobium (up to 10 fold excess over Mo). Other alloying elements commonly found in steels, for example Ni, Co, Cr, V, Cu, etc., do not interfere up to a 500 fold excess. Details of the procedure follow.

A *standard molybdenum solution* is prepared by dissolving exactly 0.750 g pure molybdenum(VI) oxide, MoO_3, in 5 mL 0.1 M sodium hydroxide, NaOH. After dissolving carefully acidify the sample by adding 10 mL, 9 M sulfuric acid, H_2SO_4, and 10 mL 7.5 M nitric acid, HNO_3. Dilute the sample to one liter in a volumetric flask. Exactly 10.00 mL of this solution is placed in a 500 mL volumetric flask and diluted with water to its final volume. This diluted solution contains 10.0 μg Mo/mL.

A *calibration curve* is prepared by placing into individual 125 mL separatory funnels between 2.00 and 15.0 mL of diluted molybdenum standard. This corresponds to a range of molybdenum values from 20.0 μg to 150 μg. To each funnel add 15 mL 5.0 M H_2SO_4 and 0.20 mL iron(III) chloride (containing 50 mg Fe(III) per mL dissolved in 6 M hydrochloric acid, HCl). Add 5 mL 10% (w/v) ascorbic acid (containing 10 g $C_6H_8O_6$, 2 mL 0.1 M Na_2-H_2EDTA, and 10 drops 98% formic acid, HCOOH, per 100 ml of solution). Add 0.2 mL 15% (w/v) titanium(III) chloride, $TiCl_3$. Mix the contents thoroughly and add 3 mL 25% (w/v) ammonium thiocyanate, NH_4SCN. (Prepare fresh daily.) Mix and allow each sample to stand 20 min. Add, to each sample, 1 mL 0.025 M tetraphenylarsonium chloride, $(C_6H_5)_4AsCl$. Now add 10 mL chloroform, $CHCl_3$. (The chloroform solution contains 20 mL 1% (w/v) hydroquinone in ethanol per 250 mL of $CHCl_3$. Prepare fresh daily). Shake each funnel for 30 s and after the layers have separated, collect the chloroform layer in a 25 mL volumetric flask by passing it from the funnel through filter paper (Whatman number 1 or equivalent). Repeat extraction of the aqueous phase with an additional 5 mL of treated chloroform. Combine the extracts and dilute each to exactly 25 mL with additional hydroquinone containing chloroform. Absorbance of each standard is measured at 470 nm in one centimeter cells against a reagent blank prepared as described but omitting molybdenum. Prepare a calibration curve of absorbance reading vs. molybdenum content.

An appropriate aliquot of the molybdenum containing unknown in acid solution is placed in a 125 mL separatory funnel and sufficient 5.0 M H_2SO_4 added to give a final acid content of about two molar. If the sample already contains iron(III), it is not necessary to add additional iron(III). However, the amount of ascorbic acid may have to be increased especially for low molybdenum concentrations. Five mL of ascorbic acid for each 100 mg Fe(III) is satisfactory. Proceed with addition of other reagents as described in the preparation of molybdenum standards. Concentration of molybdenum in the unknown aliquot is read from the calibration curve and corrected for dilution to find the amount of molybdenum present in the original sample.

1.2 Toluene-3,4-dithiol

Toluene-3,4-dithiol (4-methyl-1,2-dimercaptobenzene) is widely used for colorimetric determination of molybdenum and other metals. In addition to being unstable, the reagent is also toxic [34]. It is necessary, therefore, to use extreme caution when working with this substance. Molybdenum(VI) forms a green complex with dithiol in strong acid solution which when extracted into an organic solvent, for example butyl acetate, exhibits an absorbance maximum near 670 nm and has a molar absorptivity of about $(1.4 \text{ to } 1.8) \times 10^4$ L mol^{-1} cm^{-1} [35, 36]. Interference from tungsten, iron(III), and copper is minimized by adding respectively citric acid, ascorbic acid, and iodide ion [37]. Thiourea also masks copper [38, 39]. For solutions 8 M in hydrochloric acid the tin dithiol complex is unstable thus eliminating its interference [39].

Following is the procedure of De Silva for determination of molybdenum with dithiol [36]. It incorporates a preliminary extraction step with tributyl phosphate to eliminate many possible interfering ions and employs a glacial acetic acid-tributyl phosphate-phosphoric acid medium for optimum results. Wavelength of maximum absorbance with this procedure is 705 nm. Molar absorptivity is 18,400 L mol^{-1} cm^{-1}. A linear relation between absorbance and molybdenum concentration, Beer's law, although not reported by the author can be expected to hold from about 0.4 to 4 µg Mo/mL. Absorbance readings are stable for 24 h. With the preliminary tributyl phosphate extraction interference from the following ions in ten fold excess over molybdenum is considered negligible: Pb, Cd, Cu, Co, Ni, Zn, As, Bi, Sn, Fe, and W. Details of the procedure follow.

A standard molybdenum solution is prepared by dissolving 0.252 g sodium molybdate, $Na_2MoO_4 \cdot 2H_2O$, in distilled water and diluting to exactly one liter in a volumetric flask. Exactly 10.00 ml of this solution is placed in a 50 mL volumetric flask and glacial acetic acid, $HC_2H_3O_2$, added until exactly 50 mL of solution is present. This sample contains 20.0 µg Mo/mL. Transfer between 1.00 und 5.00 mL (20 to 100 µg Mo) of the diluted stock solution to separate 50 mL volumetric flasks. To each flask add 5 mL tributyl phosphate, $[CH_3(CH_2)_3O]_3PO$, (saturated with 6 M hydrochloric acid, HCl), 20 mL glacial acetic acid, and 2 mL 15 M phosphoric acid, H_3PO_4. Carefully add to each flask 5 drops 1% (w/v) toluene-3,4-dithiol. (Prepare by dissolving 1 g $(HS)_2C_6H_3CH_3$ in 100 mL 10% (w/v) sodium hydroxide, NaOH, solution containing 1.0 g mercaptoacetic acid, $HSCH_2COOH$. Store in a refrigerator. Discard the solution upon the first visible sign of deterioration, generally after a few weeks. Dithiol is a *toxic* substance. Handle it carefully.) Mix the contents of each flask and dilute each to exactly 50 mL with glacial acetic acid. Mix again. The flasks are set at room temperature for 3 h before proceeding. Measure the absorbance of each standard at 705 nm in a one centimeter cell against a reagent blank which substitutes 5 mL of glacial acetic acid for the molybdenum stock solution. The blank solution is treated as the standards but placed in the reference beam of the spectrophotometer. Prepare a calibration curve of absorbance vs. amount of molybdenum present in each of the standards.

An aliquot of a molybdenum containing unknown, representing between 50 and 200 µg molybdenum, is acidified with 25 mL of 6 M HCl and placed into a 125 mL separatory funnel. Five mL of tributyl phosphate (saturated with 6 M HCl) is added and the funnel contents slowly mixed for five minutes. After the layers have separated, the nonaqueous layer is removed and retained. The sample layer is subjected to a second extraction with an additional 5 mL of HCl saturated tributyl phosphate. Following separation of the layers, the tributyl phosphate portions are combined and adjusted, if necessary, to a final volume of ten mL with additional tributyl phosphate. A 5.00 mL aliquot of the tributyl phosphate extract is placed in a 50 mL volumetric flask. Add to the flask 25 mL glacial acetic acid and 2 mL 15 M H_3PO_4. Add 5 drops 1% (w/v) toluene-3,4-dithiol. Mix the contents of the flask and add additional glacial acetic acid until the total volume is 50 mL. Mix the sample and allow it to set at room temperature for three hours. Measure the absorbance at 705 nm against the reagent blank used with the molybdenum standards. Determine the amount of molybdenum present in the aliquot from the calibration curve and correct this value for the various dilutions employed to find the molybdenum content of the unknown sample.

1.3 Other Color Forming Agents

Many, many color forming reagents combine with molybdenum producing complexes suitable for analytical applications. Some of the more recently reported of these are listed in Table 6-2.

Table 6-2. Complex Forming Agents for Colorimetric Determination of Molybdenum

Reagent	Condition	Wavelength of Maximum Absorbance, nm Molar Absorptivity, L mol^{-1} cm^{-1} Specific Absorptivity,[a] mL g^{-1} cm^{-1}	Concentration Range, µg Mo mL^{-1}	Ref.
alizarine red S, 0.002 M aqueous	pH 3.8, acetate	480 9.2×10^3 0.097	0.5–5.0	[40]
alizarine red S, 0.0025 M aqueous + zephiramine, 0.0025 M aqueous	pH 5.2, acetate, extract 1,2-dichloroethane	575 – –	0.1–2.5	[41]
benzohydroxamic acid, 0.1 M aqueous	pH 2–3, HCl, extract hexanol	370 3.2×10^3 0.023	5–20	[42]
benzoylacetone, 0.15 M CHCl$_3$-n-BuOH 1:1	pH 2–3, HCl, extract	372 3.9×10^3 0.041	1–20	[43]
N-benzylaniline, 5% (w/v) CHCl$_3$	0.5 M HCl, thioglycolic acid, extract	358 3.0×10^3 0.031	5–25	[44]

Table 6-2 (continued)

Reagent	Condition	Wavelength of Maximum Absorbance, nm Molar Absorptivity, L mol^{-1} cm^{-1} Specific Absorptivity,[a] mL g^{-1} cm^{-1}	Concentration Range, μg Mo mL^{-1}	Ref.
benzyldimethylaniline HCl, 0.1 M aqueous	pH 4, acetate, pyrogallol red, extract BuOH	582 2.1×10^4 0.22	0.7–6.0	[45]
bromopyrgallol red, 0.001 M 50% (v/v) EtOH + cetylpyridine HCl, 0.01 M aqueous	0.05 M H$_2$SO$_4$, ascorbic acid	630 8.0×10^4 0.83	0.08–1.4	[46]
carminic acid, 0.005 M acetate buffer	pH 4–5, acetate	565 1.4×10^4 0.14	1.5–8.0	[47]
2,2'-dihydroxybenzophenone thiosemicarbazone, 0.2% (w/v) EtOH	0.2 M HCl, Sn(II), wait 8 min	500 3.3×10^3 0.034	1–22	[48]
2,3-dihydroxynaphthalene-6-sulfonate Na$^+$, 1% (w/v) aqueous	pH 5.2, acetate	420 7×10^3 0.07	4–20	[49]
1,4-dihydroxyphthalimide dithiosemicarbazone 0.03% (w/v) DMF	pH 2.6, ClCH$_2$COOH-NaOH, extract i-pentanol	435 9.4×10^3 0.098	10–95	[50]
ethyl xanthate K$^+$, 1% (w/v) aqueous	3 M HCl, extract acetophenone	380 6.0×10^4 0.62	1.2–13.8	[51]
flavon-3-ol-2'-sulfonic acid, 0.02 M aqueous	0.3 M HClO$_4$	370 1.4×10^4 0.15	0.09–8.20	[52]
o-hydroxyhydroquinone-phthalein, 0.001 M aqueous	pH 1.8, HCl, LT-221 surfactant	520 1.3×10^5 1.36	0.01–1.0	[53]
4-hydroxy-3-mercaptocoumarin, 0.5% (w/v) aqueous	1 M HCl, extract CHCl$_3$	500 1.2×10^4 0.12	1–7	[54]
3-hydroxy-2-methyl-1-phenyl-4-pyridone, 0.002 M CHCl$_3$	1 M HCl, extract	317 2.5×10^4 0.26	0.5–5.0	[55]
8-hydroxyquinoline, 0.6% (w/v) acetic acid	0.04 M H$_2$SO$_4$, hydrazine, 40% (v/v) pyridine	405 4.4×10^3 0.046	0.8–24	[56]
2-mercapto-3-(2-furyl)-propenoic acid, 1% (w/v) EtOH	pH 1, HCl, wait 5 min extract CHCl$_3$	450 8.6×10^3 0.090	0.5–22	[57]

Table 6-2 (continued)

Reagent	Condition	Wavelength of Maximum Absorbance, nm Molar Absorptivity, L mol^{-1} cm^{-1} Specific Absorptivity,[a] mL g^{-1} cm^{-1}	Concentration Range, μg Mo mL^{-1}	Ref.
N-methoxy-N-p-tolylcinamo-hydroxamic acid, 0.01 M CHCl$_3$	6 M HCl, extract	390 1.1×10^5 1.15	0.05–1.0	[58]
o-nitrophenylfluorone, 0.003 M EtOH + dianti-pyrinylmethane, 0.1 M CHCl$_3$	1 M HCl, extract CHCl$_3$	530 1.3×10^5 1.36	0.07–0.6	[59]
4-(phenylazo)catechol, 0.01 M benzene	pH 1.3, HCl	550 1.1×10^5 1.12	0.1–7.7	[60]
phenylfluorone, 0.001 M MeOH + cetylpyridine HCl, 0.01 M aqueous	pH 1.5, HCl, 45 °C for 30 min	540 9.6×10^3 0.10	4–60	[61]
phenylhydrazine sulfinic acid, 4% (w/v) aqueous	7 M HC$_2$H$_3$O$_2$, ascorbic acid	510 8.0×10^3 0.083	0.6–6.0	[62]
pyrocatechol, 0.3 M aqueous + triphenyltetrazolium Cl$^-$, 0,4 M aqueous	pH 0.5, HCl + KCl, extract C$_2$H$_2$CL$_2$	640 – –	2.5–15	[63]
pyrocatechol violet, 0.002 M aqueous	pH 2.7, chloroacetic acid	540 2.7×10^4 0.28	0.4–4	[40]
pyrogallol red, 0.005 M 50% (v/v) EtOH + cetyltri-methylammonium Br, 0.003 M aqueous	pH 3.6, acetate, ascorbic acid, wait 10 min	600 9.0×10^4 0.94	0.1–0.4	[64]
1-pyrrolidinecarbodithioate NH$_4^+$ salt, 0.01 M aqueous	pH 4.8, acetate, wait 40 min, extract CHCl$_3$	380 4.2×10^3 0.044	4–18	[65]
quinalizarin, 1% (w/v) EtOH	50% (v/v) EtOH, pH 5.5, acetate, 30 °C for 30 min	540 9.8×10^3 0.10	3.5–21	[66]
rubeanic acid, 0.5% (v/v) EtOH	6 M H$_2$SO$_4$, 2-propanol, dimethylformamide	600 2.2×10^3 0.023	6–22	[67]
rutin, 0.001 M 50% (v/v) MeOH	pH 3.8, formate	400 2.2×10^4 0.23	0.9–3.5	[68]

[a] Specific absorptivity corresponds to absorbance that would be observed from a solution containing 1 μg Mo/mL when measured in a cell of one centimeter path length

Table 6-3. Colorimetric Determination of Molybdenum in Specific Materials

Material	Reagent	Reference
rocks and minerals		
phosphate rocks	thiocyanate	[69]
silicate rocks	thiocyanate	[70]
silicate rocks	toluene-3,4-dithiol	[71]
silicate rocks	salicylhydroxamic acid	[72]
mordenite	5,7-dibromo-8-hydroxyquinoline	[73]
molybdenite	8-hydroxyquinoline	[56]
molybdenite	pyrogallol red + cetrimide	[74]
scheelite	thiocyanate	[75]
Cu-Pb ore	toluene-3,4-dithiol	[37]
Nb ore	thiocyanate	[76]
U ore	thiocyanate	[77]
W ore	toluene-3,4-dithiol	[78]
ore	hydroxyamidine	[79]
ore	thiocyanate + N-hydroxy-N-p-chlorophenyl-N'- (2-methyl-4-chlorophenyl)benzamidine	[80]
ore	catechol	[81]
ferrous alloys		
cast iron	thiocyanate	[82, 83]
ferromolybdenum	2',3',4'-trihydroxypropiophenone	[84]
ferromolybdenum	3,3',4,4',5-pentahydroxybenzophenone	[85]
ferromolybdenum	sulfonitrazo	[86]
ferromolybdenum	4-butyrylpyrogallol	[87]
Cr-Al steel	rezarson	[88, 89]
Cr-Mo steel	phenylhydrazine	[90]
Cr-Mo steel	2-amino-4-chlorobenzenethiol HCl	[91]
Cr-Mo steel	stilbazo + zephiramine	[92]
Cr-Mo steel	thiocyanate	[10]
Cr-Mo steel	1-nitroso-2-naphthol	[93]
Cr-Ni-Mo steel	phenylhydrazine	[94]
Cr-Ni-Mo steel	thiocyanate	[11]
Cr-V steel	dihydroquercetin	[95]
Cr-V-W steel	thiocyanate	[96]
Cr-V-W steel	mercaptoacetic acid	[97]
stainless steel	salicylaldehyde thiosemicarbazone	[98]
stainless steel	mercaptoacetic acid	[99]
Ni steel	thiocyanate + tricaprylylmethylammonium Cl	[3]
Ni steel	thiocyanate + hexamethyl phosphortriamide	[100]
W steel	thiocyanate	[101]
W steel	toluene-3,4-dithiol	[102]
low alloy steel	phenylfluorone	[103]
low alloy steel	carminic acid	[104]
low alloy steel	thiocyanate	[105]
low alloy steel	2-hydroxy-1-naphthaldehyde- isonicotinoylhydrazone	[106]
high alloy steel	thiocyanate	[107]
high alloy steel	morpholinium-4-morpholinecarbodithioate	[108]
high alloy steel	malenonitrile dithiolate	[109]

Table 6-3 (continued)

Material	Reagent	Reference
high alloy steel	2,6,7-trihydroxy-9- (4-nitrophenyl)xanthen-3-one	[110]
high alloy steel	thiocyanate + ethyl xanthate	[111]
high alloy steel	thiocyanate + 2-mercaptobenzo-γ-thiopyrone	[112]
tool steel	2-aminobenzenethiol	[113]
steel	N,N-diethylaniline	[114]
steel	thiocyanate	[115]
steel	2'-bromo-4',5'-dihydroxy- azobenzene-4'-sulfonic acid	[116]
steel	catechol + butyltriphenylphosphonium Br	[117]
non ferrous alloys		
Ni alloy	bromopyrogallol red	[118]
Ni alloy	5,7-dibromo-8-quinolinol	[119]
Nb alloy	thiocyanate	[120, 121]
Ta alloy	thiomalic acid	[122]
Ti alloy	chloro complex, ultraviolet	[123]
U alloy	hydrogen peroxide	[124]
U alloy	8-hydroxyquinoline	[125]
W alloy	carboxygallanilide	[126]
Zr alloy	sulfochlorophenol S	[127]
pure elements		
beryllium	thiocyanate	[128]
niobium	thiocyanate + crystal violet	[18, 129]
plutonium	toluene-3,4-dithiol	[130]
rhenium	thiocyanate + crystal violet	[18]
tantalum	toluene-3,4-dithiol	[131]
titanium	salicylfluorene	[132]
tungsten	thiocyanate + crystal violet	[18]
tungsten	4-(2-pyridylazo)-resorcinol + hydroxylamine	[133]
uranium	8-hydroxyquinoline	[134]
vanadium	thiocyanate	[135]
yttrium	thiocyanate	[136]
zirconium	thiocyanate	[137]
biological samples		
human blood	benzohydroxamine acid	[138, 139]
urine	toluene-3,4-dithiol	[140]
feces	thiocyanate	[141]
cadaver tissue	thiocyanate	[142]
sheep liver	toluene-3,4-dithiol	[143]
milk	toluene-3,4-dithiol	[144]
alfalfa	thiocyanate	[145]
barley	toluene-3,4-dithiol	[146]
clover	toluene-3,4-dithiol	[37]
grasses	thiocyanate	[147]
oats	2-amino-4-chloro-benzenethiol	[148]
soybean	bromopyrogallol red + hexadecylpyridinium chloride	[46]

Table 6-3 (continued)

Material	Reagent	Reference
tomato	thiocyanate	[149]
soil	thiocyanate	[150]
soil	toluene-3,4-dithiol	[151]
soil	thiocyanate + rhodamine B	[152]
soil	2′,3′,4′,trihydroxyazobenzene-4′-sulfonic acid	[153]
coal	thiocyanate	[154]
oil shale	thiocyanate	[155]
environmental samples		
natural water	pyrogallol red + dimethyldiocta-decylammonium ion	[156]
natural water	4-amino-5-hydroxynaphthalene-2,7-disulfonic acid	[157]
sea water	chloro-8-hydroxy-7-iodoquinoline	[158]
waste water	pyrogallol red + hydrogen peroxide	[159]
salt brine	thiocyanate	[160]
atmospheric particulates	thiocyanate	[161]
miscellaneous samples		
drugs	toluene-3,4-dithiol	[162]
whiskey	thiocyanate	[163]

2. Applications to Specific Materials

Colorimetric procedures have traditionally been used for estimating trace amounts of a component. Although replaced, to some extent, by the more recent techniques of atomic absorption spectrometry and voltammetry, they still play an important and continuing role in analysis. The availability of instrumentation, development of new color forming reagents, and familiarity of the procedure by many analysts, all contribute to selection of colorimetric procedures over other available approaches. Molybdenum in almost every conceivable matrix has been, at one time or another, subjected to colorimetric determination. In some instances, familiar reagents are employed to form the necessary colored complex. In other instances, more exotic reagents are used. Table 6-3 lists many materials analyzed for molybdenum by colorimetric procedures.

References

1. Hiskey, C.F., Meloche, V.W.: J. Am. Chem. Soc., *62*, 1565 (1940)
2. Ul'ko, N.V.: Ukr. Khim. Zh., *31*, 887 (1965)
3. Wilson, A.M., McFarland, O.K.: Anal. Chem., *36*, 2488 (1964)
4. Mitchell, P.C.H., Williams, R.J.P.: J. Chem. Soc., 4570 (1962)
5. Sasaki, Y., Taylor, R.S., Sykes, A.G.: J. Chem. Soc. Dalton Trans. 396 (1975)

6. Greenland, L.P., Lillie, E.G.: Anal. Chim. Acta, *69*, 335 (1974)
7. Yatirajam, V., Ram, J.: Mikrochim. Acta, 671 (1974)
8. Bergh, A.A., Haught, G.P.: Inorg. Chem., *8*, 189 (1969)
9. Crouthamel, C.E., Johnson, C.E.: Anal. Chem., *26*, 1284 (1954)
10. Bermejo-Martínez, F., Latas-Pérez, P.: Fresenius' Z. Anal. Chem., *264*, 139 (1973)
11. Luke, C.L.: Anal. Chim. Acta, *34*, 302 (1966)
12. Burriel-Marti, F., Cobo, G.A.: An. Quim., *64*, 973 (1968)
13. Lazarev, A.I., Lazareva, V.I.: Zavod. Lab., *24*, 798 (1958)
14. Podberazskaya, N.K.: Tr. Kaz. Nauchno-Issled. Inst. Miner. Syr'ya, (2), 232 (1960); Chem. Abstr., *60*, 2321d (1964)
15. Marov, I.N., Dubrov, Yu.N., Ermakov, A.N., Martyinova, G.N.: Zh. Neorg. Khim., *13*, 1674 (1968)
16. Debal, E., Chassin, R., Peynot, S.: Talanta, *23*, 35 (1976)
17. Markle, G.E., Boltz, D.F.: Anal. Chem., *25*, 1261 (1953)
18. Solomatin, V.T., Yakovlev, P.Ya., Lapshina, L.A., Artemova, T.N.: Zh. Anal. Khim., *30*, 114 (1975)
19. Neeb, R.: Fresenius' Z. Anal. Chem., *182*, 10 (1961)
20. Peng, P.Y., Sandell, E.B.: Anal. Chim. Acta, *29*, 325 (1963)
21. Potrokhov, V.K., Lebedeva, L.I.: Zh. Anal. Khim., *21*, 182 (1966)
22. Burriel-Marti, F., Bouza, A.P.: Quim. Ind. (Bilbao), *3*, 168 (1956); Chem. Abstr., *53*, 21414i (1959)
23. Bergamin Filho, H., Medeiros, J.X., Reis, B.F., Zagatto, E.A.G.: Anal. Chim. Acta, *101*, 9 (1978)
24. Pilipenko, A.T., Solomeina, Z.G.: Ukr. Khim. Zh., *39*, 1169 (1973)
25. Puzanowska-Tarasiewicz, H., Grudniewska, A., Tarasiewicz, M.: Anal. Chim. Acta, *94*, 435 (1977)
26. Ganago, L.I., Ivanova, I.F.: Zh. Anal. Khim., *27*, 713 (1972)
27. Verma, R.M., Patel, K.S., Mishra, R.K.: Fresenius' Z. Anal. Chem., *307*, 128 (1981)
28. Arunachalam, M.K., Kumaran, M.K.: Talanta, *21*, 355 (1974)
29. Nytko, K., Sikorska-Tomicka, H.: Microchem. J. *25*, 548 (1980)
30. Koralewski, T.J., Parker, G.A.: Anal. Chim. Acta, *113*, 389 (1980)
31. Gowda, H.S., Ramappa, P.G., Manjappa, S.: Indian J. Chem., *18A*, 276 (1979)
32. Fogg, A.G., Kumar, J.L., Burns, D.T.: Analyst (London), *100*, 311 (1975)
33. Rao, T.P., Ramakrishna, T.V.: Bull. Chem. Soc. Jpn., *53*, 2380 (1980)
34. Christensen, H.E., Luginbyhl, T.T., Carroll, B.S.: The Toxic Substance List 1974 Edition, Publication No. NIOSH 74-134, Rockville, Maryland, U.S. Department of Health, Education and Welfare, 1974
35. Gilbert, T.W., Sandell, E.B.: J. Am. Chem. Soc., *82*, 1087 (1960)
36. De Silva, M.E.M.: Analyst (London), *100*, 517 (1975)
37. Quin, B.F., Brooks, R.R.: Anal. Chim. Acta, *74*, 75 (1975)
38. Bingley, J.B.: J. Agric. Food Chem., *11*, 130 (1963)
39. Milham, P.J., Maksvytis, A., Barkus, B.: Anal. Chem., *44*, 2102 (1972)
40. Csányi, L.J.: Mikrochim. Acta, I, 277 (1980)
41. Ishibashi, N., Kohara, H., Abe, K.: Bunseki Kagaku, *17*, 154 (1968)
42. Agrawal, Y.K., Maru, P.C., Sharma, T.P., Patke, S., Verma, P.C.: Fresenius' Z. Anal. Chem., *276*, 300 (1975)
43. Mit'kina, L.I., Mel'chakova, N.V., Peshkova, V.M.: Zh. Anal. Khim., *36*(6), 1099 (1981)
44. Rao, S.P., Bhargava, R.N., Reddy, R.R.: Indian J. Chem., *20A*, 639 (1981)
45. Postoronko, A.I., Golosnitskaya, V.A.: Zh. Neorg. Khim., *25*(9), 2446 (1980)
46. Savvin, S.B., Chernova, R.K., Beloliptseva, C.M.: Zh. Anal. Khim., *35*, 1128 (1980)
47. Lee, A.P., Boltz, D.F.: Microchem. J., *17*, 380 (1972)

48. Lopez Fernandez, J.M., Perez-Bendito, D., Valcarcel, M.: Analyst (London), *103*, 1210 (1978)
49. Buchwald, H., Richardson, E.: Talanta, *9*, 631 (1962)
50. Radriguez, M.T.: Analyst (London), *107*, 41 (1982)
51. Arunachalam, M.K., Kumaran, M.K.: Talanta, *21*, 355 (1974)
52. Yamamoto, K., Hara, J., Ohashi, K.: Anal. Chim. Acta, *135*, 173 (1982)
53. Mori, I., Fujita, Y., Kamata, Y., Enoki, T.: Bunseki Kagaku, *27*, 259 (1978)
54. Bag, S.P., Chaktrabarti, A.K., Goswami, J.P.: J. Indian Chem. Soc., *58*, 901 (1981)
55. Tamhina, B., Herak, M.J.: Mikrochim. Acta, I, 47 (1977)
56. Nishida, H.: Bunseki Kagaku, *22*, 771 (1973)
57. Izquierdo, A., Calmet, J.: Analusis, *4*, 200 (1976)
58. Agrawal, Y.K., Jain, R.K.: Croat. Chem. Acta, *54*, 249 (1981)
59. Ganago, L.I., Ivanova, I.F.: Zh. Anal. Khim., *35*, 1138 (1980)
60. Wakamatsu, Y.: Bunseki Kagaku, *29*, 472 (1980)
61. Mori, I., Yamamoto, S., Enoki, T.: Bunseki Kagaku, *22*, 1061 (1973)
62. Jain, S.K., Pandey, D.C., Joseph, K.C., Satyanarayana, K.: J. Indian Chem. Soc., *56*, 353 (1979)
63. Alexandrov, A., Kostova, M.: Mikrochim. Acta, I 487 (1976)
64. Wyganowski, C.: Microchem. J., *25*, 147 (1980)
65. Lee, A.P., Boltz, D.F.: Anal. Lett., *8*, 345 (1975)
66. Rani, S., Banerji, S.K.: Microchem. J., *18*, 636 (1973)
67. Williams, D.A., Holcomb, I.J., Boltz, D.F.: Anal. Chem., *47*, 2025 (1975)
68. Pfendt, L.B., Krunz, M.M., Janjić, T.J.: Mikrochim. Acta, I, 385 (1980)
69. Robinson, W.O.: Soil Sci., *66*, 317 (1948)
70. Lillie, E.G., Greenland, L.P.: Anal. Chim. Acta, *69*, 313 (1974)
71. Kawabuchi, K., Kuroda, R.: Talanta, *17*, 67 (1970)
72. Pal, C.K., Chakraburtty, A.K.: J. Indian Chem. Soc., *52*, 138 (1975)
73. Olaru, M., Basceanu, M.: Rev. Chim. (Bucharest), *31*, 598 (1980); Chem. Abstr., *93*, 197102j (1980)
74. Liu, S., Xu, Q., Wang, Z.: Fen Hsi Hua Hsueh, *9*, 311 (1981); Anal. Abstr., *42*, 2B147 (1982)
75. Hope, R.P.: Anal. Chem., *29*, 1053 (1957)
76. Sen, S.: Sci. Cult., *23*, 318 (1957)
77. Korkisch, J., Steffan, I.: Mikrochim. Acta, 651 (1973)
78. Jeffrey, P.G.: Analyst (London), *82*, 558 (1957)
79. Kharsan, R.S., Mishra, R.K.: Bull. Chem. Soc. Jpn., *53*, 1736 (1980)
80. Kharsan, R.S., Patel, K.S., Deb, K.K., Mishra, R.K.: Fresenius' Z. Anal. Chem., *295*, 415 (1979)
81. Patrovský, V.: Chem. Listy, *49*, 854 (1955)
82. Patil, S.P., Shinde, V.M.: Anal. Chim. Acta, *67*, 473 (1973)
83. Szocs, E., Szocs, S.: Rev. Chim. (Bucharest), *27*, 713 (1976); Chem. Abstr., *86*, 83192p (1977)
84. Popa, G., Dumitrescu, V.: Revta Chim., *26*, 761 (1975); Anal. Abstr., *30*, 4B117 (1976)
85. Popa, G., Dumitrescu, V., Dumitrescu, N. Vegh, B.: Rev. Chim. (Bucharest), 26, 160 (1975); Chem. Abstr., *83*, 107785a (1975)
86. Rybina, T.F.: Zavod. Lab., *41*, 797 (1975)
87. Dumitrescu, V., Vegh, B.: Rev. Chim. (Bucharest), *28*, 268 (1977); Chem. Abstr., *88*, 57899w (1978)
88. Shevko, V.A., Eremenko, S.N., Frumina, N.S.: Zavod. Lab., *41*, 1061 (1975)
89. Lukin, A.M., Petrova, G.S., Kaslina, N.A.: Zh. Anal. Khim., *24*, 39 (1969)
90. Mutte, P.A., Mannur, M.: Chim. Anal., *38*, 94 (1956)
91. Kirkbright, G.F., Yoe, J.H.: Talanta, *11*, 415 (1964)

 92. Ozawa, T., Okutani, T., Utsumi, S.: Bunseki Kagaku, *22*, 1592 (1973)
 93. Peshkova, V.M., Ivanova, E.K., Memon, S.: Zh. Anal. Khim., *35*, 486 (1980)
 94. Ayres, G.H., Tuffly, B.L.: Anal. Chem., *23*, 304 (1951)
 95. Chan, F.L., Moshier, R.W.: Talanta, *3*, 272 (1960)
 96. Bauer, G.A.: Anal. Chem., *37*, 155 (1965)
 97. Otterson, D.A., Graab, J.W.: Anal. Chem., *30*, 1282 (1958)
 98. Perez-Bendito, D., Pino-Perez, F.: Mikrochim. Acta, 613 (1976)
 99. Will, F., Yoe, J.H.: Anal. Chem., *25*, 1363 (1953)
100. Mitra, M., Mitra, B.K.: Talanta, *24*, 698 (1977)
101. Uehara, F., Furano, A.: Bunseki Kagaku, *19*, 183 (1970)
102. Wells, J.E., Pemberton, R.: Analyst (London), *72*, 185 (1947)
103. Black, A.H., Bonfiglio, J.D.: Anal. Chem., *33*, 431 (1961)
104. Kirkbright, G.F., West, T.S., Woodward, C.: Talanta, *13*, 1645 (1966)
105. Braithwaite, K., Hobson, J.D.: Analyst (London), *93*, 633 (1968)
106. Podchainova, V.N., Yatsenko, V.A., Dolgorev, A.V.: Zavod, Lab., *40*, 243 (1974)
107. Lounamaa, N.: Anal. Chim. Acta, *33*, 21 (1965)
108. Likussar, W., Beyer, W., Wawschinek, O.: Mikrochim. Acta, 735 (1968)
109. Chakrabarti, A.K., Bag, S.P.: Talanta, *19*, 1187 (1972)
110. Bagdasarov, K.N., Shchemeleva, G.G., Mikalauskas, T.: Zavod. Lab., *39*, 1047 (1973)
111. Arunachalam, M.K., Kumaran, M.K.: Talanta, *21*, 355 (1974)
112. Savariar, C.P., Arunachalam, M.K., Hariharan, T.R.: Anal. Chim. Acta, *69*, 305 (1974)
113. Chakrabarti, A.K., Bag, S.P.: Talanta, *23*, 736 (1976)
114. Babendo, A.S., Mineeva, L.A., Godovskaya, K.I.: Zavod. Lab., *42*, 1035 (1976)
115. Gorlach, V.F., Marchenko, L.M.: Ukr. Khim. Zh., *40*, 983 (1974)
116. Wakamatsu, Y.: Bunseki Kagaku, *26*, 470 (1977)
117. Docekalova-Klapalova, H., Vrchlabsky, M.: Hutn. Lisky, *31*, 356 (1976); Chem. Abstr., *86*, 11474j (1977)
118. Andreeva, I.Yu., Golubtsova, Z.G., Lebedeva, L.I., Sukhorukova, N.P.: Stand. Obraztsy Chern. Metall., (9), 48 (1980); Anal. Abstr., *41*, 1B146 (1981)
119. Gorina, D.O., Zharovskii, F.G.: Zavod. Lab., *35*, 794 (1969)
120. Privalova, M.M., Tulina, M.D.: Zavod. Lab., *33*(1), 16 (1967)
121. Solomatin, V.T., Lapshina, L.A., Artemova, T.N., Yakovlev, P.Ya.: Zavod. Lab., *40*, 937 (1974)
122. Polyak, L.Ya., Bashkirova, I.S.: Zh. Anal. Khim., *22*(2), 200 (1967)
123. Engel, O.: Hutn. Listy, *28*, 816 (1973); Chem. Abstr., *80*, 127845m (1974)
124. Bacon, A., Milner, G.W.C.: Anal. Chim. Acta, *15*, 573 (1956)
125. Mutojima, K., Hashitani, H.: Anal. Chem., *33*, 48 (1961)
126. Pashchenko, E.N., Mal'tsev, V.F., Zherikova, A.N.: Zh. Anal. Khim., *31*, 400 (1976)
127. Elinson, S.V., Savvin, S.B., Nezhnova, T.I.: Zh. Anal. Khim., *22*, 531 (1967)
128. Hibbits, J.O., Davis, W.F., Menke, M.R.: Talanta, *4*, 104 (1960)·
129. Ivanova, I.F., Ganago, L.I., Semenovich, I.A.: Izv. Vyssh. Uchebn. Zavod., Khim. Khim. Tekhnol., *20*, 1815 (1977); Chem. Abstr., *88*, 114797n (1978)
130. Nelson, G.B., Waterbury, G.R.: U S At. Energy Comm., T1D-7629 (1961); Chem. Abstr., *57*, 9204a (1962)
131. Kallmann, S., Hobart, E.W., Oberthin, H.K.: Talanta, *15*, 982 (1968)
132. Shustova, M.B., Nazarenko, V.A.: Khim. Prom. Nauk.-Tekhnol. Zb., (4), 78 (1962); Chem. Abstr., *59*, 5762b (1963)
133. Barkovskii, V.F., Guseva, L.N.: Tr. Inst. Khim. Ural. Nauch. Tsentr. Akad. Nauk SSSR, (27), 108 (1974); Anal. Abstr., *29*, 5B203 (1975)
134. Motojina, K., Hashitani, H., Izawa, K., Yoshidu, H.: Bunseki Kagaku, *11*, 47 (1962)
135. Klug, O.N., Metlenko, S.: Chem. Anal. (Warsaw), *13*, 7 (1968)

136. Hibbits, J.O., Davis, W.F., Menke, M.R.: US At. Energy Comm., APEX-519 (1959); Chem. Abstr., *54*, 13969c (1960)
137. Goward, G.W., Burd, R.M.: US At. Energy Comm., WAPD-CTA (GLA)-192 (1957); Chem. Abstr., *53*, 13881i (1959)
138. Agrawal, Y.K.: J. Indian Chem. Soc., *54*, 451 (1977)
139. Agrawal, Y.K.: Anal. Lett., *13B*, 357 (1980)
140. Galli, A.: Ann. Biol. Clin. Paris, *24*, 165 (1966); Chem. Abstr., *64*, 18009f (1966)
141. Tompseff, S.L., Fitzpatric, J.: Analyst (London), *75*, 279 (1950)
142. Rubtsov, A.F.: Vopr. Sud. Med. Ekspertizy, Sb. (Ashkhabad: Turkmenistan) No. 1, 127 (1965); Chem. Abstr., *65*, 17349d (1966)
143. Norhein, G., Waasjoe, E.: Fresenius' Z. Anal. Chem., *286*, 229 (1977)
144. Stanton, R.E., Hardwick, A.J.: Analyst (London), *93*, 193 (1968)
145. Stanfield, K.E.: Ind. Eng. Chem., Anal. Ed., *7*, 273 (1935)
146. Heanes, D.L.: Analyst (London), *106*, 172 (1981)
147. Carel, A.B., Wimberley, J.W.: Anal. Lett. *15A*, 493 (1982)
148. Ssekaalo, H., Johnson, R.M.: J. Sci. Food Agr., *20*, 581 (1969)
149. Johnson, C.M., Arkley, T.H.: Anal. Chem., *26*, 572 (1954)
150. Grigg, J.L.: Analyst (London), *78*, 470 (1953)
151. Stanton, R.E., Mockler, M., Newton, S.: J. Geochem. Explor., *2*, 37 (1973)
152. Haddad, P.R., Alexander, P.W., Smythe, L.E.: Talanta, *22*, 61 (1975)
153. Gambarov, D.G., Guseinov, A.G.: Zh. Anal. Khim., *33*, 1343 (1978)
154. Monkovskii, M.A., Gordon, S.A., Nurminskii, N.N.: Zavod. Lab., *28*, 1321 (1962)
155. Vahemets, H.: Uch. Zap. Tartu. Gos. Univ. No. 193, 121 (1966); Chem. Abstr., *69*, 40971y (1968)
156. Morgen, E.A., Rossinskaya, E.S., Vlasov, N.A.: Zh. Anal. Khim., *30*, 1384 (1975)
157. Lavrelashvili, L.V., Vasnev, A.N., Dzotsenidze, N.E., Kreingol'd, S.U.: Izv. Akad. Nauk Gruz. SSR, Ser. Khim., 6(1), 20 (1980); Anal. Abstr., *40*, 3H69 (1981)
158. Ohta, N., Fujita, M., Tomura, K.: Bunseki Kagaku, *28*, 277 (1979)
159. Mori, T., Fujita, Y., Sakaguchi, K., Enoki, T.: Bunseki Kagaku, *29*, 413 (1980)
160. Gürtler, O.: Fresenius' Z. Anal. Chem., *285*, 259 (1977)
161. Tabor, E.C.: Health Lab. Sci., *7*, 149 (1970)
162. Bickford, C.F., Jones, W.S., Keene, J.S.: J. Am. Pharm. Assoc., Sci. Ed., *37*, 255 (1948)
163. Pro, M.J., Nelson, R.A.: J. Assoc. Off. Agr. Chem., *39*, 945 (1956)

Chapter 7

Emission Spectroscopy

With the development of new and more effective excitation sources, emission spectroscopy is again becoming one of the major analytical techniques for examining complex materials, both qualitatively and quantitatively. The inductively coupled plasma, ICP, source in particular, with its extremely high operating temperature, has established new and lower limits of detection for many elements [1, 2]. Detection limits for molybdenum using conventional and plasma sources are:

dc arc source [3]	1 µg/g	(313.259 nm)
	1 µg/g	(317.035 nm)
plasma jet [4a]	0.1 µg/mL	(281.615 nm)
radiofrequency plasma torch [4b]	0.045 µg/mL	(379.825 nm)
inductively coupled plasma [4c]	0.0002 µg/mL	(379.825 nm)

Not only metal alloys, traditional samples for spectroscopic analysis, but biological and environmental samples, with minute amounts of molybdenum, are now studied by emission spectroscopy.

1. General Considerations

Listings of spectral lines for molybdenum are available as part of exhaustive compilations of atomic spectra [5–9]. The emissions commonly used in analytical investigations of molybdenum are presented in Table 7-1 [10, 11].

Of the lines listed in Table 7-1 the 379.8, 317.0, and 313.3 nm emissions are perhaps most often employed when determining molybdenum. The 386.4 nm line is preferred by some in steel analysis because of overlap at 379.8 and 317.0 nm from iron lines [12]. Others [13] recommend the 202.0 nm line for molybdenum in steels. With steels containing higher amounts of molybdenum (5 to 10%) the 550.6 and 553.3 nm lines, because of their lower sensitivities, are preferred [14]. Either the 317.0 nm [15] or 379.8 nm [16] emission is used for molybdenum present in soil samples and the 317.0 nm emission for molybdenum in rocks [17, 18]. In all cases, other investigators have selected alternate emissions for molybdenum in these and other matrices. There is no consensus regarding which atomic emission of molybdenum is best for a particular application and the reader attempting to determine molybdenum for the first time may wish to repeat his/her measurements several times focusing each time upon a different molybdenum emission to ascertain the best conditions for a particular sample. In addition to overlap from iron emissions,

Table 7-1. Atomic Emission Lines of Molybdenum Used in Chemical Analysis

	Wavelength nm	Energy eV
II	202.032	6.13
II	203.846	6.08
II	281.615	6.06
I	313.259	3.96
I	315.817	3.93
I	317.035	3.91
I	379.825	3.26
I	386.411	3.20
I	390.296	3.17
I	423.259	5.01
I	427.724	4.43
I	481.925	5.20
I	536.056	5.58
I	550.649	3.58
I	553.305	3.57

other elements, too, exhibit spectra that overlap selected molybdenum lines. These interferences are too numerous to catalog here and one should consult appropriate reference listings [19–21].

Not only overlap but also enhancement or decreased intensity of molybdenum emission occurs depending upon the presence of other elements in the sample matrix. Varying acid strength of liquid samples is, to some degree, responsible for changes in intensity of molybdenum emissions [22]. This is especially important with plasma sources which generally require liquid samples. Nitric, sulfuric, and hydrochloric acid all decrease emission intensity. Best results for steels when using an ultra high frequency plasma torch were obtained when samples and standards contain 0.5 M hydrochloric acid [22]. Large amounts of sodium ion decrease molybdenum emission [23]. Sodium chloride is sometimes added to both samples and standards so the effect will be uniform for both. Large amounts of potassium also effect molybdenum emissions [24]. Various alkali salts [25] and alkaline earth compounds [26] also interfere. Studies have been made upon the matrix effects of animal tissue [27], plants [28], rocks and minerals [29], and lake sediments [30] when analyzing for trace elements including molybdenum. Frequently an element known to enhance the spectral emissions of molybdenum or to counter the decrease in intensity from other matrix elements is intentionally added to sample and standards with the hope of increasing sensitivity. Aluminum salts added to soil samples help overcome errors due to calcium [16]. Improved volatility of molybdenum halides prompts one to add either fluoride or chloride compounds to sample mixtures. Silver chloride [31, 32], copper(I) chloride [33], sodium chloride [34], copper(II) fluoride [35], trifluoromethane [36], and polytetrafluoroethylene, PTFE [37] are some of the compounds added. Solid molyb-

Table 7-2. Spectroscopic Buffers Used in the Determination of Molybdenum

Buffer Composition	Buffer Sample Ratio	Application	Ref.
AgCl	1:1	niobium oxides	[38]
CuCl	1:1	silicon oxides	[33]
In_2O_3	5:1	airborne particulates	[39]
SiO_2:K_2SO_4 (3:2)	1:1	soils	[28]
Al_2O_3:graphite (1:1)	1:1	mineral waters	[40]
CaF_2:graphite (1:2)	1:1	lake sediment	[30]
LiF:graphite (1:4)	3:1	rocks	[41]
PTFE:graphite (2:5)	10:7	iron	[37]
ZnO:graphite (1:1)	2:1	tantalum oxides	[42]
$CaCO_3$:Li_2CO_3:graphite (1:1:3)	5:1	plants	[43]
CuO:$SrCO_3$:graphite (6:12:31)	15:1	slag	[44]
Al_2O_3:$CaCO_3$:K_2CO_3 (7:3:1)	1:1	rocks	[18]

Table 7-3. Internal Standards for the Spectroscopic Determination of Molybdenum

Molybdenum Emission, nm	Internal Standard Emission, nm		Application	Ref.
202.032	Fe	176.138	steel	[13]
281.615	Co	269.468	steel	[45]
	Fe	279.240	iron	[37]
	Fe	282.328	steel	[46]
	Fe	282.863	steel	[47]
	Zr	294.894	Zr alloy	[48]
313.259	Co	304.400	plants	[49]
	Cu	305.120	copper	[50]
	Mg	292.875	magnesium	[51]
	Pd	324.270	plutonium	[52]
	Pt	311.013	platinum	[53]
315.817	Ni	323.465	ore concentrates	[54]
317.035	Co	304.400	airborne particles	[55]
	Cu	282.437	slag	[44]
	Cu	301.084	W alloy	[56]

denum containing samples are frequently mixed with a spectroscopic buffer, a mixture of appropriate salts present in large amount to assure a uniform matrix and perhaps to enhance the molybdenum signal. Various buffer mixtures are common in spectroscopic studies. In simultaneous multielement determinations they are frequently chosen to give the best overall results for several trace components. Table 7-2 lists some buffer mixtures used in the spectroscopic determination of molybdenum.

It is customary to employ an *internal standard* in spectroscopic analysis. The internal standard present in a fixed amount in all known and unknown samples serves to compensate for fluctuations in excitation conditions and/or variations in photographic development. Intensity ratios of analyte emission to internal standard emission rather than intensity of analyte emission alone are used in determining the concentration of desired constituent. Frequently choice of an internal standard is governed by the type of sample under study. Sometimes the background signal is used as internal standard. Whatever internal standard is finally selected, it should have an emission near that of the analyte and ideally behave in a manner similar to the analyte when experimental conditions are varied. Some internal standards for molybdenum are listed in Table 7-3.

Sample concentration is sometimes employed prior to spectroscopic measurement especially when only minute amounts of molybdenum are present in a sample. This is especially true for samples of natural waters where the concentration of molybdenum at best is minimal. Co-precipitation [57], adsorption on carbon in the presence of organic precipitating agents [58, 59], and extraction [60] all serve to concentrate molybdenum from natural waters before spectroscopic examination. Molybdenum is concentrated from biological samples by precipitation as the mixed thiocyanate chromopyrazole I complex [61].

2. Laboratory Procedures

Defavault and Marshall [18] have developed a procedure for determination of μg quantities of molybdenum in various rock samples. Using dc arc excitation and measurement at 317.0347 nm satisfactory results are obtained from a series of U.S. Geological Survey and Geological Survey of Canada standard samples. Iron emission at 317.0346 nm was not a problem with samples containing only a few percent iron. The 316.9854 calcium line, also, did not interfere. Following are details of their procedure.

Operating Conditions		
Electrodes	sample (anode)	graphite rode ¼ in. outside diameter, 1½ in. long, drilled $^3/_{16}$ in. deep by $^{11}/_{64}$ in. diameter (National SPK 3829)
	counter (cathode)	graphite rod shaped to a point (National L 3803)
Excitation		dc arc 110 v stabilized to about 30 v between electrodes, 6–7 amp for first 10 sec increasing to 15 amp for 35 sec additional
Spectrograph		Hilger large quartz E-742
Wavelength range		245.0–350.0 nm
Slit width		15 μ, 20% neutral density filter on arc stand
Exposure time		45 sec total exposure
Photographic plate		Kodak spectrum analyzer No. 1
Photographic processing		develop 5 min at 22 °C with Kodak HRP developer
Densitometry		ARL spectroline scanner, zero setting from plate near analytic line

A series of standards is prepared in an artificial rock base. Prepare the synthetic matrix by thoroughly grinding together the following high purity chemicals:

315 g	silicon dioxide, SiO_2
100 g	aluminum oxide, Al_2O_3
25 g	iron(III) oxide, Fe_2O_3
17.5 g	calcium carbonate, $CaCo_3$
10.0 g	magnesium oxide, MgO
17.5 g	sodium carbonate, Na_2CO_3
17.5 g	potassium carbonate, K_2CO_3

Add exactly 7.5 mg molybdenum oxide, MoO_3, to 500 g of the synthetic rock mixture. Thoroughly grind the sample until the molybdenum is uniformly distributed throughout the matrix and the matrix passes through an 80 mesh screen. This sample contains 10.0 µg Mo/g mixture. A series of molybdenum standards is prepared containing from 0.1 µg Mo/g to 10.0 µg Mo/g by diluting appropriate amounts of this mixture with graphite. The 0.1 µg Mo/g standard, for example, contains 1.0 g of the prepared molybdenum matrix added to 99 g of graphite.

For each standard, combine 150 mg of sample with 150 mg of Tennant's buffer (consisting of 7.0 g aluminum oxide, 3.0 g calcium carbonate, and 1.0 g potassium carbonate ground to 80 mesh or finer). One to two drops of 25% (w/v) sucrose, $C_{12}H_{22}O_{11}$ (in water-ethanol, 1:3) is added during mixing. Approximately half of the mixture is used to fill each electrode. The sample, 150 mg in the form of a fine powder (80 mesh or finer), is also prepared in this manner.

The prepared electrodes, containing standards and samples, are placed in an aluminum rack and dried upright on a hot plate at 300 °C for one-half hour. After drying, record the emission spectrum of each. Only those runs for which the residue within the electrode, after arcing, appears as a free bead are considered valid. The authors find that those samples which have spread on the walls of the electrode cup result in low values for molybdenum and in high background emissions. Once the spectra are obtained, the plates are developed and densitometer readings taken. The data are treated in the usual manner for spectroscopic determinations. Those unfamiliar with these procedures should consult appropriate references [62, 63].

Rock samples analyzed by the authors contained molybdenum within the standard range of 0.1 to 10.0 µg Mo/g sample.

Operating Conditions

Instrumentation	Hitachi u.h.f. plasma torch spectrometer, model 300
For the plasma torch	frequency; 2450 Mhz
	flow rate plasma forming gas (Ar); 3.0 mL/min
	flow rate plasma sheath gas (Ar); 3,5 mL/min
	field current; 370 mA
	anode current; 270 mA
Slit width	entrance 30 µm
	exit 50 µm

Akatsuka and Atsuya [12] have used an *ultra high frequency plasma torch* as excitation source for determining molybdenum in steels. Measurements are made upon the 386.411 nm emission and other elements generally present in steels do not seriously interfere. Iron, to some degree suppresses matrix interferences while enhancing molybdenum emissions. It is added to the molybdenum standards, bringing them more in line with the sample matrix. All solutions aspirated into the plasma are adjusted to 0.5 M HCl for optimum results. Following are details of their procedure.

Exactly one gram pure molybdenum metal is dissolved in a small volume of aqua regia, HCl-HNO$_3$ 3:1 (v/v). The solution is evaporated to near dryness and the residue dissolved in 100 mL 6.0 M hydrochloric acid, HCl. The solution is diluted to exactly one liter with

Table 7-4. Determination of Molybdenum in Specific Materials by Atomic Emission Spectroscopy

Material	Reference
rocks and minerals	
andesite	[18]
besalt	[64]
granite	[65]
quartz	[29]
igneous rocks	[66]
silicate rocks	[67]
molybdenite	[68]
Au ore	[69]
Cu ore	[70]
Fe ore	[71]
Sb ore	[72]
ore concentrate	[54]
ferrous alloys	
ferromolybdenum	[23]
slag	[44]
stainless steel	[73]
Nb steel	[74]
low alloy steel	[13, 75]
high alloy steel	[45, 47, 76]
steel	[12, 14, 46, 77, 78, 79]
non ferrous alloys	
Co alloy	[80]
Mg alloy	[51]
Ni alloy	[81]
Ti alloy	[82, 83]
U alloy	[84, 85]
W-Re alloy	[56]
Zr alloy	[86, 87]
pure elements	
beryllium	[88]
bismuth	[89, 90]

Table 7-4. (continued)

Material	Reference
boron	[91]
carbon	[92]
copper	[50]
gallium	[93]
iron	[37]
lead	[94]
platinum	[53]
plutonium	[52]
silicon	[95]
thorium	[96]
tin	[97]
tungsten	[98]
uranium	[99, 100]
yttrium	[101]
zirconium	[48]
metal oxides	
Nb_2O_5	[38]
rare earth oxides	[102, 103]
SiO_2	[33]
Ta_2O_5	[42]
ThO_2	[104]
TiO_2	[105]
U_3O_8	[106]
WO_3	[98]
Y_2O_3	[101]
ZrO_2	[31]
biological samples	
animal tissue	[27, 107]
blood	[108]
feces	[109]
alfalfa	[49]
brocolli	[110]
peas	[111]
tomato	[112]
wheat	[113]
plants	[28, 43, 114, 115]
soils	[15, 16, 35, 116, 117, 118, 119]
coal	[17]
lubricating oils	[120]
environmental samples	
natural water	[57, 58, 59]
sea water	[60]
mineral water	[40]
waste water	[121]
sediments	[30, 122]
airborne particulates	[39, 55]
glass	[123]

distilled water. This solution contains 1.00 mg Mo/mL. A series of molybdenum standards is prepared containing from 0 to 50 µg Mo/mL by placing the appropriate amount of molybdenum stock solution in separate 100 mL volumetric flasks. For the 50 µg Mo/mL solution, place 5.00 mL of stock solution in a 100 mL volumetric flask. If X equals the volume of molybdenum stock solution added to each flask, add also (5.00 − X) mL 0.6 M HCl to each flask to maintain constant acidity for every standard. Add to each standard 25.0 mL of iron(III) solution (prepared by dissolving 20.0 g pure iron metal in 350 mL 12 M HCl and diluting the solution to exactly 500 mL with distilled water). Add sufficient distilled water until each standard is in a final volume of exactly 100 mL. Each standard contains 10 mg Fe/mL. The unknown steel sample, between 0.5 and 1.5 g depending upon the molybdenum content, is dissolved in 15 mL aqua regia. The solution is evaporated to near dryness and the residue dissolved in 20 mL of 2.5 M HCl. The sample is transferred to a 100 mL volumetric flask and distilled water added until the final volume is 100 mL. The emission spectrum of each standard and sample is recorded and the data obtained processed in the usual manner to find the percent molybdenum in the steel.

3. Applications to Specific Materials

Because of its multielement capability, emission spectroscopy is often used for examining minor and trace components in various materials. Table 7-4 cites some of these applications for molybdenum.

References

1. Van Montfort, P.F.E., Agterdenbos, J.: Talanta, *28*, 629 (1981)
2. Boumans, P.W.J.M.: Mikrochim. Acta, I, 399 (1978)
3. Addink, N.W.H.: Spectrochim. Acta, *11*, 168 (1957)
4a. Greenfield, S., McGeachin, H.M., Smith, P.B.: Talanta, *22*, 1 (1975)
4b. Ibid.: *22*, 553 (1975)
4c. Ibid.: *23*, 1 (1976)
5. Harrison, G.R.: Massachusetts Institute of Technology Wavelength Tables, 1969 Cambridge, MA, Massachusetts Institute of Technology
6. Zaidel', A.N., Prokof'ev, V.K., Raiskii, S.M., Slavnyi, V.A., Shreider, E.Ya.: Tables of Spectral Lines, 1970, New York, IFI/Plenum
7. Schauls, M.R., Sawyer, R.A.: Phys. Rev., *58*, 78 (1940)
8. Winge, R.K., Peterson, V.J., Fassel, V.A.: Appl. Spectrosc., *33*, 206 (1979)
9. Boumans, P.W.J.M.: Spectrochim. Acta, *36B*, 169 (1981)
10. Török, T., Mika, J., Gegus, E.: Emission Spectrochemical Analysis 1978 Bristol, Adam Hilger
11. Dean, J.A.: Lange's Handbook of Chemistry, 12th Edition, 1979, New York, McGraw-Hill
12. Akatsuka, K., Atsuya, I.: Anal. Chim. Acta, *99*, 351 (1978)
13. Goto, H., Saito, A.: Bunseki Kagaku, *17*, 194 (1968)
14. Novotny, M., Oulehla, K.: Hutn. Listy, *24*, 535 (1969)
15. McKenzie, R.M.: Spectrochim. Acta, *18*, 1009 (1962)
16. Manzoori, J.L.: Talanta, *27*, 682 (1980)
17. Swaine, D.J.: Spectrochim. Acta, *19*, 841 (1963)
18. Delavault, R.E., Marshall, D.B.: Can. J. Spectrosc., *18*, 10 (1973)

19. Kroonen, J., Vader, D.: Line Interference in Emission Spectrographic Analysis, 1963, Amsterdam, Elsevier
20. Parsons, M.L., Forster, A., Anderson, D.: An Atlas of Spectral Interference in ICP Spectroscopy, 1980, New York, Plenum
21. Boumans, P.W.J.M.: Line Coincidence Tables for Inductively Coupled Plasma Atomic Emission Spectrometry, 1981, Oxford, Pergamon Press
22. Goryaenov, N.N., Serebrovskii, V.I.: Zavod. Lab., *44*, 693 (1978)
23. Suzuki, M.: Bunseki Kagaku, *19*, 207 (1970)
24. Woodis, T.C., Holmes, J.H., Ardis, J.D., Johnson, F.J.: J. Assoc. Off. Anal. Chem., *63*, 1245 (1980)
25. Kozhevnikov, L.A., Shavrin, A.M.: Uch. Zap. Perm. Gos. Univ., *19*, 129 (1960); Chem. Abstr., *58*, 5029h (1963)
26. Burriel-Marti, F., Jimenez-Gomez, S.: An. Real Soc. Espan. Fis. Quim., Ser. B, *48*, 783 (1952); Chem. Abstr., *47*, 4243c (1953)
27. Neidermeier, W., Griggs, J.H., Webb, J.: Appl. Spectrosc., *28*, 1 (1974)
28. Snopov, N.G.: Zh. Prikl. Spektrosk., *9*, 919 (1968)
29. Ivanova, G.F.: Zh. Anal. Khim., *21*, 1307 (1966)
30. Cowgill, U.M.: Appl. Spectrosc., *28*, 455 (1974)
31. Nakajima, T., Fukushima, H.: Bunseki Kagaku, *9*, 81 (1960)
32. Degtyareva, O., Sinitsyna, L.G.: Zh. Anal. Khim., *20*, 603 (1965)
33. Shuvalova, E.I., Khlystova, A.D., Tarasevich, N.I., Gongalyuk, N.G.: Zh. Anal. Khim., *29*, 384 (1974)
34. Mohamed, M., Tarasevich, N.I.: Vestn. Mosk. Univ., Ser. II, *21*, 92 (1966); Chem. Abstr., *66*, 24095y (1967)
35. Pometun, E.A., Sinyakova, V.G.: Zh. Anal. Khim., *22*, 440 (1967)
36. Kirkbright, G.F., Snook, R.D.: Anal. Chem., *51*, 1938 (1979)
37. Morello, R., De Gregorio, P., Savastano, G.: Appl. Spectrosc., *28*, 14 (1974)
38. Spitz, J., Chazee, J.J., Trans Van Danh: Method. Phys. Anal., *4*, 375 (1968)
39. Iami, S., Kusaka, Y., Tsuji, H., Hrshiya, Y.: Anal. Chim. Acta, *108*, 103 (1979)
40. Pepin, D., Gardes, A., Petit, J., Berger, J.A., Gaillard, G.: Analusis, *2*, 549 (1973)
41. Watson, A.E., Russell, G.M.: Rep. Natn. Inst. Metall., (1656), 1 (1974); Anal. Abstr., *29*, 4B9 (1975)
42. Chandola, L.C., Dixit, V.S.: Curr. Sci., *43*, 372 (1974)
43. Gabriels, R., Cottenie, A.: Lab. Pract., *25*, 835 (1976)
44. Lapinskaya, M.E., Kuznetsova, G.V., Shilt, E.S.: Zavod. Lab., *40*, 1478 (1974)
45. Waggoner, C.A.: Appl. Spectrosc., *13*, 31 (1959)
46. Monteaux Corréa de Sá, Y.: An. Assoc. Quim. Brasil, *9*, 59 (1950); Chem. Abstr., *46*, 376h (1952)
47. Gerasimova, S.I.: Tr. Vses. Nauchn.-Issled. Konstruckt. Inst. Khim. Mashinostr., (45), 155 (1963); Chem. Abstr., *62*, 1073b (1965)
48. Farrell, R.F., Harter, G.J., Jacobs, R.M.: Anal. Chem., *31*, 1550 (1959)
49. Mathis, W.T.: Anal. Chem., *25*, 943 (1953)
50. Publicover, W.: Anal. Chem., *37*, 1680 (1965)
51. Apsher, A.D.: Appl. Spectrosc., *32*, 212 (1978)
52. Dhumwad, R.K., Joshi, M.V., Patwardham, A.B.: Anal. Chim. Acta, *42*, 334 (1968)
53. Lincoln, A.J., Kohler, J.C.: Anal. Chem., *34*, 1247 (1962)
54. Trapitsyn, N.F., Mullayanov, F.I.: Zavod. Lab., *28*, 950 (1962)
55. Sugimae, A.: Appl. Spectrosc., *28*, 458 (1974)
56. Tarasevich, N.I., Chebotarev, V.E.: Zh. Anal. Khim., *28*, 1023 (1973)
57. Lebedinskaya, M.P., Chuiko, V.T.: Zh. Anal. Khim., *28*, 2413 (1973)
58. Korganova, T.S., Polyakov, V.A., Kolotov, B.A., Nechaeva, T.P.: Zavod. Lab., *39*, 1186 (1973)

59. Zaguzin, V.P., Ksenzova, V.I., Pogrebyak, Yu.F.: Zh. Anal. Khim., *35*, 1143 (1980)
60. McLeod, C.W., Otsuki, A., Okamoto, K., Haraguchi, H., Fuwa, K.: Analyst (London), *106*, 419 (1981)
61. Chelnokova, M.N., Kosareva, T.M., Kurmaeva, T.V.: Agrokhimiya, (1), 129 (1979); Chem. Abstr., *90*, 134669s (1979)
62. Brode, W.R.: Chemical Spectroscopy, 1943, New York, John Wiley
63. Nachtrieb, N.H.: Principles and Practice of Spectrochemical Analysis, 1950, New York, McGraw-Hill
64. Chan, K.M., Riley, J.P.: Chem. Geol., *2*, 171 (1967)
65. Ivanova, G.F.: Zh. Anal. Khim., *20*, 82 (1965)
66. Pavelenko, L.I.: Zh. Anal. Khim., *15*, 463 (1960)
67. Spackov, A.: Collect. Czeh. Chem. Commun., *30*, 1255 (1965)
68. Nedler, V.V.: Zavod. Lab., *7*, 795 (1938)
69. Nedler, V.V.: Zavod. Lab., *7*, 57 (1938)
70. Breckenridge, R.L., Russell, G.M., Watson, A.G.: Rep. Natn. Inst. Metall., (1783), 1 (1976); Anal. Abstr., *32*, 6B13 (1977)
71. Kreshkov, A.P., Kuchkarev, E.A., Davydova, T.Ya.: Tr. Mosk. Khim.-Tekhnol. Inst. No.*85*, 185 (1975); Anal. Abstr., *31*, 5B149 (1976)
72. Abashidze, N.F.: Tr. Kavk. Inst. Miner. Syr'ya No.*1*, 137 (1960); Chem. Abstr., *58*, 3876b (1963)
73. Goto, H., Kakita, Y., Namiki, M.: Bunseki Kagaku, *19*, 1467 (1970)
74. Suzuki, M.: Bunseki Kagaku, *18*, 176 (1969)
75. Butler, C.C., Kniseley, R.N., Fassel, V.A.: Anal. Chem., *47*, 825 (1975)
76. Sherstyuk, A.I., Vovk, V.N., Vlasova, L.I.: Zavod. Lab., *40*, 1471 (1974)
77. Burriel-Marti, F., Alvarez Herrero, C.: Quim. Anal., *28*, 15 (1974)
78. Endo, Y., Sakao, N.: Bunseki Kagaku, *30*, 433 (1981)
79. Thompson, M., Goulter, J.E.: Analyst (London), *106*, 32 (1981)
80. Sakai, S., Yoshino, K.: Bunko Kenkyu, *10*, 23 (1961); Chem. Abstr., *57*, 15787d (1962)
81. Geber, W.O., Tobin, W.H.: Appl. Spectrosc., *8*, 120 (1954)
82. Gusarskii, V.V., Frieman, G.I., Borisenko, V.F., Medvedeva, M.S.: Zh. Anal. Khim., *31*, 2187 (1976)
83. Iwasaki, K., Uchida, H., Tanaka, K.: Anal. Chim. Acta, *135*, 369 (1982)
84. Purcell, D.H.: Appl. Spectrosc., *15*, 101 (1961)
85. Roca, M.: An. Real Soc. Espan. Lis. Quim., *62B*, 1165 (1966); Chem. Abstr., *66*, 111228n (1967)
86. Frain, J.F., Ryan, J.R., Jacobs, R.M.: US At. Energy Comm., WAPD-CTA (GLA)-162-7 (1958); Chem. Abstr., *54*, 20625g (1960)
87. Nishizaki, K., Matsumura, T., Miwa, H.: Nippon Kinzoku Gakkaishi, *30*, 94 (1966); Chem. Abstr., *65*, 6285g (1966)
88. Carpenter, L., Lewis, R.W., Hazen, K.A.: Appl. Spectrosc., *20*, 44 (1966)
89. Forrest, J., Finston, H.L.: Appl. Spectrosc., *14*, 127 (1960)
90. Baranova, L.L., Solodovnik, S.M., Blokh, I.M.: Zh. Anal. Khim., *28*, 1417 (1973)
91. Degtyareva, O.F., Ostrovskaya, M.F.: Zh. Anal. Khim., *22*, 1863 (1967)
92. Webb, M.S.W., Cotterill, J.C., Jones, T.W.: Anal. Chim. Acta, *26*, 548 (1962)
93. Oldfield, J.H., Bridge, E.P.: Analyst (London), *86*, 267 (1961)
94. Karabash, A.G., Bondarenko, L.S., Morozova, G.G., Peizulaev, Sh.I.: Zh. Anal. Khim., *15*, 623 (1960)
95. Vivarat-Perrin, J., Bonnier, E.: Chim. Anal. (Paris), *48*, 511 (1966)
96. Degtyareva, O.F., Sinitsyna, L.G.: Zh. Anal. Khim., *22*, 1500 (1967)
97. Malakhov, U.V., Protopopova, N.P., Trukhacheva, V.A., Yudelevich, I.G.: Tr. Kom. Anal. Khim. Akad. Nauk SSSR, Inst. Geokhim. Anal. Khim., *16*, 89 (1968); Chem. Abstr., *69*, 24279y (1968)

98. Harizanov, Yu., Jordanov, N.: Talanta, *22*, 485 (1975)
99. Dixit, R.M., Venkatasubramanian, R., Khanna, P.P., Sindgikar, K.S.B.: Rep. Bhabha At. Res. Centre, BARC-664 (1973); Anal. Abstr., *27*, 1277 (1974)
100. Moseeva, E.P., Pinchuk, G.P., Sokolov, A.B., Karabash, A.G., Peizulaev, Sh.I.: Zh. Anal. Khim., *29*, 1589 (1974)
101. Slyusareva, R.L., Kondrat'eva, L.I., Peizulaev, Sh.I.: Zavod. Lab., *39*, 1465 (1973)
102. Kaplan, B.Ya., Kirillova, Z.P., Merisov, Yu.L., Nazarova, M.G., Petrova, E.I., Skripkin, G.S.: Zavod. Lab., *40*(3), 256 (1974)
103. Mishchenko, V.T., Koval'chuk, L.I., Ponomarenko, L.P., Smirnova, L.V.: Ukr. Khim. Zh., *47*(7), 767 (1981)
104. Nakajima, T., Fukushima, H.: Bunseki Kagaku, *9*, 830 (1960)
105. Zahhariya, N.F., Turulina, O.P.: Zh. Anal. Khim., *29*, 1170 (1974)
106. Avni, R.: Spectrochim. Acta, *23B*, 619 (1968)
107. Schramel, P., Klose, B.-J., Hasse, S.: Fresenius' Z. Anal. Chem., *310*, 209 (1982)
108. Granovskii, E.I., Aksinenko, I.A.: Tr. Inst. Kraev. Patol. Kaz. SSR, (22), 245 (1971); Chem. Abstr., *79*, 15392j (1973)
109. Fisher, M.T., Lee, J.: Anal. Chim. Acta, *139*, 333 (1982)
110. Lyons, D.J., Roofayel, R.L.: Analyst (London), *107*, 331 (1982)
111. Ghosh, M.K., Mazumder, K.C.: Indian J. Phys., *24*, 67 (1950)
112. Ecrement, F., Burelli, F.P.: Analusis, *2*, 306 (1973)
113. Chelnokova, M.N., Busev, A.I., Kosareva, T.M.: Zh. Anal. Khim., *36*, 230 (1981)
114. Oertel, A.C.: Aust. J. Appl. Sci., *1*, 259 (1950)
115. Jones, J.W., Capar, S.G., O'Haver, T.C.: Analyst (London), *107*, 353 (1982)
116. Bol'shakov, V.A., Dobritskaya, Yu.I., Ivanov, D.N., Orlova, L.P.: Agrokhimiya, (1), 142 (1967); Chem. Abstr., *66*, 121778f (1967)
117. Manzoori, J.L.: Talanta, *27*, 682 (1980)
118. Savinova, E.N., Korobova, E.M., Shumskaya, T.V.: Zh. Anal. Khim., *36*(7), 1267 (1981)
119. Pavlenko, L.I., Karyakin, A.V., Berti, F.: Zh. Anal. Khim., *36*(9), 1793 (1981)
120. Kauffman, R.E., Saba, C.S., Rhine, W.E., Eisentraut, K.J.: Anal. Chem., *54*, 975 (1982)
121. Broekaert, J.A.C., Leis, F.: Anal. Chim. Acta, *109*, 73 (1979)
122. Oedegard, M.: J. Geochem. Explor., *14*, 119 (1981)
123. Floyd, M.A., Fassel, V.A., Winge, B.K., Katzenberger, J.M., D'Silva, A.P.: Anal. Chem., *52*, 431 (1980)

Chapter 8

Atomic Absorption Spectrometry

Atomic absorption spectrometry, a technique whereby the absorbance of radiation from an external source by free individual atoms of analyte depends upon analyte concentration, is a sensitive and selective analytical procedure. It finds widespread use, especially when trace amounts of material are present and other procedures would be incapable of generating sufficient response. Sensitivity for molybdenum by atomic absorption spectrometry is not as great as for some other metals because of the refractory nature of many molybdenum compounds. With special precautions, however, and/or preliminary concentration steps determination of molybdenum at the µg/mL level and below is common place.

1. General Considerations

Before molybdenum in an unknown sample can absorb radiation from an external molybdenum source, the chemical bonds joining molybdenum to itself and to other constituents within the sample must be broken, that is, the sample must be atomized. This requires considerable energy, especially for molybdenum. Both flame and non flame sources are used to supply this energy and a variety of fuel mixtures, burner types, etc. are available. In general, air-acetylene and nitrous oxide-acetylene flames are used, usually with a fuel rich flame. The detection limits for many elements, both theoretical and experimental, using these two flame mixtures and atomic absorption measurements, have been compiled [1]. Detection limit is taken as the perceived signal twice that of the instrument noise level, that is, twice the signal observed when no analyte is present in the flame or furnace. This value varies from spectrometer to spectrometer and with variations in experimental parameters. The authors [1] report for molybdenum, monitoring the 313.26 nm line, experimental detection limits of 0.02 and 0.03 µg Mo/mL for air-acetylene and nitrous oxide-acetylene flames respectively.

More information is available on the detection limits for molybdenum in various flames when flame emission rather than atomic absorbance is measured, that is, when the emission of atomized molybdenum is detected rather than monitor the absorbance of radiation (atomic absorbance) from an external source. With flame emission the detection limit for various flames are:

oxyhydrogen flame at 379.82 nm – 2 µg Mo/mL [2]
oxyacetylene flame at 379.82 nm – 0.5 µg Mo/mL [3]
separated air acetylene flame at 379.82 nm – 2 µg Mo/mL [4]
separated nitrous oxide-acetylene flame at 313.26 nm – 0.5 µg Mo/mL [5]
nitrous oxide-acetylene flame at 390.29 nm – 0.1 µg Mo/mL [6].

Atomic fluorescence rather than atomic absorption measurement of molybdenum at 313.26 nm produces a detection limit of 0.8 µg Mo/mL with a separated nitrous oxide-acetylene flame [7].

Wavelengths most frequently monitored for molybdenum with atomic absorption measurements are 313.259 and 379.825 nm. Other lines sometimes used, to avoid interferences or because of a high molybdenum concentration a less sensitive line is desirable, are [8–10]:

315.81 nm
317.04 nm
319.39 nm
386.41 nm
390.29 nm.

An alternative to flame excitation of samples for atomic absorption spectrometry is the use of flameless, electrically heated chambers. These take the form of hollow carbon rods, refractory metal boats, and other designs. Generally, an inert gas flow encompasses the chamber to exclude oxygen. This reduces the chance of refractory oxide formation. With these devices sample size is usually a few microliters. Samples are placed in the chamber and, through controlled temperature programming, the solvent is first evaporated. For solid carbonbased materials, dry ashing is achieved. This is followed by a period of intermediate temperature during which more volatile sample components are removed. Finally, at the maximum desired temperature, the analyte is volatilized into the optical path. For molybdenum temperatures in excess of 2,100 °C are necessary for atomization with a graphite furnace [11] with temperatures upwards of 2,600 °C preferred [12]. With graphite furnaces, molybdenum carbide formation occurs decreasing sensitivity [13]. A coating of pyrolytic graphite on the carbon tube helps eliminate this problem [14, 15]. The detection limit for molybdenum when using a heated carbon filament under optimum conditions with a one µL sample is 0.03 µg Mo/mL [16].

Factors effecting the determination of molybdenum by atomic absorption spectrometry have been studied. Variations in fuel gas-oxidant ratio, for example, while generally adjusted to favor a fuel rich, reducing, flame can, if excessive fuel is present, result in incomplete combustion of carbon and production of molybdenum carbide. This will decrease sensitivity of the procedure for molybdenum [17]. It is prudent to experimentally determine the optimum gas ratio for the particular spectrometer and burner assembly used in your laboratory.

Much concern in determining molybdenum by atomic absorption spectrometry is directed towards the effect of diverse ions present with molybdenum in a sample. Measurement of absorbance for a given molybdenum concentration in the presence of any of about forty diverse ions found that each effected the absorbance reading to some degree [18]. Some ions decrease molybdenum absorbance. It is not possible to state exactly what change in signal a given diverse ion will produce because the effect varies with choice of burner gases and with concentration of the diverse ion. In general, alkaline earth metals, some refractory metals, and certain acids decrease the expected molybdenum signal although this decrease is countered, somewhat, by the presence of acetate ion [19]. This decrease is attributed in part to formation of high melting compounds, for example, alkaline earth molybdates and to heteropoly formation. Other ions enhance the absorbance signal of a given molybdenum sample. Aluminum chloride [20, 21], ammonium chloride [22], and sodium or potassium sulfate [23] all produce an absorbance signal for molybdenum greater than would be expected if these ions were absent from the sample. This enhancement is attributed to a lateral, horizontal, diffusion of molybdenum in the flame when certain ions are present, resulting in increased concentration of molybdenum in the flame center over what would otherwise be present [24, 25]. Potassium chloride, in the presence of 5-sulfosalicylic acid, with a nitrous oxide-acetylene flame, also causes enhanced molybdenum signal [26]. Perhaps least understood is the effect of various mineral acids upon the absorbance reading of a molybdenum sample. Dilli, Gawne, and Ocago [27] summarize the often conflicting results observed in earlier investigations and, in addition, present their own findings regarding suppression/enhancement by various acids. They recommend dilute nitric or dilute hydrochloric acid as the best choice for use with molybdenum when it is aspirated into a flame. In higher concentrations, above two molar, the molybdenum signal is depressed by these acids [28]. Dilute perchloric acid, about 0.01 M, is shown to have negligible effect upon molybdenum absorbance readings [29]. Sulfuric acid and phosphoric acid in dilute solutions, about 0.01 M, tend to enhance a molybdenum signal but as their concentration increases, a decrease in molybdenum absorbance is observed, perhaps because of the increased viscosity of the sample solution with increasing amounts of these acids. One should be prepared to adjust his/her molybdenum standards to the same acid content present in the unknown if any doubt persists regarding possible alteration of the molybdenum signal.

A trade off exists between the undesirable decrease in atomic absorption signal because certain ions, Fe, Ca, etc., are present and enhancement of the signal by addition of Al^{3+}, NH_4Cl, etc. David [21] recommends adjusting all solutions to contain 1,000 µg Al^{3+} and 2% (w/v) NH_4Cl to suppress interferences and enhance the absorbance. Others prefer solutions containing 2% (w/v) NH_4Cl only [30], 0.08 to 0.8 M H_3PO_4 [28], 500 µg Al^{3+}/mL and 5.8% H_3PO_4 [31] or 5% (v/v) HCl + 5% (v/v) H_3PO_4 + 2.5 mg Al^{3+}/mL [32]. In addition to compensation of interferences by proper choice of acid and/or signal enhancing ions, it is sometimes desirable, as far as possible, to duplicate the sample matrix. Iron, for example, is added to molybdenum standards when determining molybdenum in steels [33]. Should the same calibration curve be used for a variety of steels, it may also be necessary to add iron to the steel samples in the case of stainless and other high alloy steels to adjust their iron content to that of the standards [33]. Addition of many elements to molybdenum standards in amounts that duplicate a plant matrix is suggested when determining molybdenum in plant material [34].

Trace amounts of molybdenum are frequently present in samples analyzed by atomic absorption spectrometry. This is especially true of biological and environmental samples. With these and other samples containing only minute amounts of molybdenum, it is sometimes desirable to introduce a preliminary concentration step prior to aspirating molybdenum into the burner assembly. By enhancing the molybdenum concentration, one avoids instrument readings near the limit of detection of many spectrometers, thus improving the quality of the data obtained. Frequently preconcentration takes the form of an extraction. This often serves the double purpose of both concentrating the molybdenum present and simultaneous-

Table 8-1. Reagents for Extracting Molybdenum Prior to Determination by Atomic Absorption Spectrometry

Reagent	Reference
Aliquot 336, methyl isobutyl ketone	[35]
ammonium pyrrolidinedithiocarbamate, methyl isobutyl ketone	[36, 37]
ammonium pyrrolidinedithiocarbamate, n-pentyl methyl ketone	[38]
α-benzoinoxime, chloroform	[39]
8-hydroxyquinoline, methyl isobutyl ketone	[40]
thiocyanate ion, methyl isobutyl ketone	[41]
thiocyanate ion + Aliquot 336, chloroform	[42]
thiocyanate ion + Amberlite LA 1, chloroform	[43]
trioctylamine oxide, toluene	[44]

ly removing possible interfering ions. Molybdenum in the organic phase can often be aspirated directly into the optical path of the spectrometer. Several reagents used for extracting molybdenum prior to determination by atomic absorption are listed in Table 8-1.

2. Laboratory Procedures

The following procedure is that of Purushottam, Naidu, and Lal devised by them for determining *molybdenum in rocks and steels* [28]. Following dissolving of the sample phosphoric acid is added to suppress interferences from Fe(III), Ca, Mg, Mn(II), Pb, Al, and Sb(III). An air-acetylene flame is used and readings are made from the 313.3 nm molybdenum line. Other parameters include a 0.10 nm slit setting, 6 mA lamp current, and 6 mL/min aspiration rate.

A *standard molybdenum solution* is prepared by dissolving 0.252 g sodium molybdate, $Na_2MoO_4 \cdot 2H_2O$ in distilled water and diluting to exactly one liter in a volumetric flask. This solution contains 100 µg Mo/mL. A series of standards is prepared in 100 mL volumetric flasks containing from 5.00 µg Mo/mL (5.00 mL of stock solution per 100 mL) to 25.0 µg Mo/mL (25.0 mL of stock solution per 100 mL). Add to each standard and to a blank containing no molybdenum 0.5 mL of 15 M phosphoric acid, H_3PO_4. Now add sufficient distilled water to each sample to adjust the final volume to exactly 100 mL. Mix the contents of each flask. The authors used an acetylene pressure of 12 lb/in² with main cylinder pressure greater than 150 lb/in². Air pressure was 13 lb/in². Using these values as a starting point and aspirating the most concentrated standard into the flame, make whatever slight adjustments are necessary in these settings to achieve maximum absorbance reading. Adjust also for maximum effect burner height and burner angle. Once all parameters are set for achieving maximum absorbance, aspirate in turn the blank and each molybdenum standard into the flame. Record absorbance values and subtract from each the absorbance value observed from the blank solution. Prepare a calibration plot of absorbance vs. concentration, µg Mo/mL, from each molybdenum standard.

The *sample containing from 0.5 to 2.5 mg Mo* is, following appropriate treatment to place it in solution, evaporated to near dryness. Water and 0.5 mL 15 M H_3PO_4 are added to dissolve the residue. The solution is diluted to 100 mL in a volumetric flask, mixed thoroughly and aspirated into the flame. From the observed absorbance reading and with the aid of the calibration curve, the concentration of molybdenum is read from the calibration curve. Apply dilution corrections if necessary and report the amount of molybdenum present in the original sample.

Kim, Alexander, and Smythe report a procedure for *molybdenum in soils and natural waters* by first extracting molybdenum prior to measurement by atomic absorption [42, 43].

Molybdenum(V) thiocyanate complex is extracted with the quaternary amine Aliquat 336, tricaprylylmethylammonium chloride, dissolved in chloroform. The chloroform is evaporated and the residue dissolved in methyl isobutyl ketone. This solution is aspirated directly into a fuel rich nitrous oxide-acetylene flame. Measurement is made using the 313.3 nm molybdenum line. Other parameters include a 0.1 nm slit setting, 5 mA lamp current, and 2.3 mL/min aspiration rate. Rhenium is the only serious interference found for this procedure. Nitrate ion and EDTA, if present, cause slight decrease in molybdenum absorbance. Metals commonly found with molybdenum do not significantly effect the results.

A *standard molybdenum solution* is prepared by dissolving 0.184 g ammonium molybdate, $(NH_4)_6Mo_7O_{24}\cdot4H_2O$ in 1.0 M hydrochloric acid, HCl. The solution is diluted to exactly one liter in a volumetric flask with additional 1.0 M HCl. Exactly 10.0 mL of this solution is transferred to a second one liter volumetric flask and additional 1.0 M HCl added to bring the final volume to 1,000 mL. This diluted stock solution contains 1.0 µg Mo/mL.

Accurately known amounts, between 1.00 mL (1.0 µg Mo) and 10.0 mL (10.0 µg Mo) of the diluted stock solution are placed in separate 125 mL separatory funnels. Sufficient additional 1.0 M HCl is added to each funnel until the final volume in each is about 40 mL. To each funnel add 2 mL 20% (w/v) potassium thiocyanate, KSCN, and 2 mL 20% (w/v) tin(II) chloride. (Prepare $SnCl_2$ by dissolving 40 g chemically pure tin metal in 40 mL 10 M HCl. When the solution is clear, cool to room temperature and carefully add distilled water until the final volume is 200 mL.) Mix the contents of each separatory funnel and allow the solutions to stand for five minutes until reduction of molybdenum is complete. Add 10 mL 0.2% (w/v) Aliquot 336 in chloroform. Shake the flask vigorously for one minute and after the layers have separated, place the organic phase into a 25 mL weighing bottle. Extract the aqueous phase a second time with an additional 10 mL of 0.2% (w/v) Aliquot 336 solution. Combine the chloroform extracts and place the weighing bottle on a steambath to evaporate the chloroform. Following evaporation, dissolve the residue in methyl isobutyl ketone. It may be necessary to warm the mixture briefly to dissolve the solid. Transfer quantitatively each methyl isobutyl ketone solution to a separate 10 mL volumetric flask and add additional MIBK until each standard is in exactly 10 mL of solution. Each standard and a blank containing pure methyl isobutyl ketone is aspirated, in turn, into the nitrous oxide-acetylene flame. Nitrous oxide flow is about 6.5 L/min and acetylene flow is about 7 L/min. Make whatever slight adjustments from these values as is necessary to obtain maximum absorbance readings from the most concentrated molybdenum standard. Absorbance readings are plotted vs. molybdenum concentrations for each standard after correcting for the blank absorbance. Concentration range for the standards is from 0.1 µg Mo/mL (1.0 mL stock Mo solution per 10 mL MIBK) to 1.0 µg Mo/mL (10.0 mL stock Mo solution per 10 mL MIBK).

The *sample containing from 1.0 to 10.0 µg Mo* is, following appropriate treatment to place it in solution, evaporated to near dryness. The residue is dissolved in about 40 mL of 1.0 M HCl and transferred to a 125 mL separatory funnel. The double extraction described in the preceeding paragraph is carried out and following removal of the chloroform solvent methyl isobutyl ketone is added to dissolve the residue. The sample is placed in a 10 mL volumetric flask and additional MIBK added until the volume is exactly 10 mL. The solution is aspirated directly into the flame. From the observed absorbance and the calibration curve, determine the concentration of molybdenum in the MIBK fraction. Correct this value for whatever dilution was employed from the original sample and report the concentration of molybdenum in the original sample.

Table 8-2. Determination of Molybdenum in Specific Materials by Atomic Absorption Spectrometry

Material	Reference	Material	Reference
rocks and minerals		Ti alloys	[65]
phosphate rocks	[37]	U alloys	[66]
silicate rocks	[38]	biological samples	
limestone	[45]	blood	[67]
quartz	[46]	dental tissue	[68]
rocks	[35, 47]	urine	[67]
molybdenite	[48]	milk xanthine oxadise	[69]
Cu-Mo ore	[39]	alfalfa	[70]
U ore	[49]	cane sugar	[71]
ore	[32]	clover	[72]
slag	[50]	forage plants	[73]
ore tailings	[51, 52]	lucerne	[38]
ferrous alloys		orchard leaves	[34]
cast iron	[30]	plants	[74, 75, 76]
ferromolybdenum	[28]	soils	[42, 49, 77]
Cr-Ni steel	[53]	environmental samples	
Cr-W-V steel	[39]	natural water	[42, 78, 79]
Cr-Ni-Nb steel	[54]	tap water	[80]
stainless steel	[55, 56]	lake water	[40]
W steel	[12]	sea water	[81, 82, 83]
low alloy steel	[33, 57]	mineral water	[84]
high speed steel	[58]	salt brines	[37, 85]
tool steel	[59]	sewage sludge	[86]
steel	[60, 61, 62]	fly ash	[87]
non ferrous alloys		miscellaneous samples	
brass	[12]	fertilizer	[88, 89]
bronze	[12]	iron containing catalyst	[90]
Ni alloys	[63]	lubricants	[91, 92]
Nb alloys	[64]	multivitamins	[93]

3. Applications to Specific Materials

Molybdenum in a variety of materials over a wide range of concentration is determined by atomic absorption spectrometry. Some of these applications are listed in Table 8-2.

References

1. Parsons, M.L., McElfresh, P.M.: Appl. Spectrosc., *26*, 472 (1972)
2. Skogerboe, R.K., Heybey, A.T., Morrison, G.H.: Anal. Chem., *38*, 1821 (1966)
3. Fassel, V.A., Myers, R.B., Kniseley, R.N.: Spectrochim. Acta, *19*, 1187 (1963)
4. Hobbs, R.S., Kirkbright, G.F., West, T.S.: Analyst (London), *94*, 554 (1969)
5. Kirkbright, G.F., Semb, A., West, T.S.: Spectrosc. Lett., *1*, 7 (1968)

6. Pickett, E.E., Koirtyohann, S.R.: Spectrochim. Acta, *23B*, 235 (1968)
7. Dagnall, R.M., Kirkbright, G.F., West, T.S., Wood, R.: Anal. Chem., *42*, 1029 (1970)
8. Pegachev, N.A., Plastinin, V.V.: Izv. Vyssh. Uchebn. Zaved., Fiz., *13*, 145 (1970); Chem. Abstr., *73*, 61146f (1970)
9. Price, W.J.: Spectrochemical Analysis by Atomic Absorption, 1979, London, Heyden
10. Slavin, M.: Atomic Absorption Spectroscopy, 1978, New York, John Wiley
11. Sturgeon, R.E., Chakrabarti, C.L., Langford, C.H.: Anal. Chem, *48*, 1792 (1976)
12. Bodrov, N.V., Nikolaev, G.I.: Zh. Anal. Khim., *24*, 1314 (1969)
13. Sneddon, J., Ottaway, J.M., Rowston, W.R.: Analyst (London), *103*, 776 (1978)
14. Manning, D.C., Ediger, R.D.: At. Absorpt. Newsl., *15*, 42 (1976)
15. Slavin, S.: At. Absorpt. Newsl., *15*, 97 (1976)
16. Johnson, D.J., West, T.S., Dagnall, R.M.: Anal. Chim. Acta, *66*, 171 (1973)
17. Sturgeon, R.E., Chakrabarti, C.L.: Anal. Chem., *48*, 677 (1976)
18. Ramakrishna, T.V., West, P.W., Robinson, J.W.: Anal. Chim. Acta, *44*, 437 (1969)
19. Tominaga, M., Umezaki, Y.: Anal. Chim. Acta, *139*, 279 (1982)
20. David, D.J.: Analyst (London), *86*, 730 (1961)
21. David, D.J.: Analyst (London), *93*, 79 (1968)
22. Edgar, R.M.: At. Absorpt. Newsl., *14*, 68 (1975)
23. Kerbyson, J.D., Ratzkowski, C.: Can. Spectrosc., *15*, 43 (1970)
24. West, A.C., Kniseley, R.N., Fassel, V.A.: Anal. Chem., *45*, 815 (1973)
25. West, A.C., Fassel, V.A., Kniseley, R.N.: Anal. Chem., *45*, 1586 (1973)
26. Komarek, J., Mahr, V., Sommer, L.: Colloct. Czech. Chem. Commun., *46*, 708 (1981)
27. Dilli, S., Gawne, K.M., Ocago, G.W.: Anal. Chim. Acta, *69*, 287 (1974)
28. Purushottam, A., Naidu, P.P., Lal, S.S.: Talanta, *19*, 1193 (1972)
29. Studnicki, M.: Anal. Chem., *51*, 1336 (1979)
30. Mostyn, R.A., Cunningham, A.F.: Anal. Chem., *38*, 121 (1966)
31. Panteleeva, E.Yu., Masalovich, N.S., Ostroumov, G.V., Polikarova, N.V.: Zh. Anal. Khim., *35*, 1885 (1980)
32. Pereverzova, E.F.: Zh. Anal. Khim., *33*, 1576 (1978)
33. Thomerson, D.R., Price, W.J.: Analyst (London), *96*, 321 (1971)
34. Wilson, D.O.: Commun. Soil Sci. Plant Anal., *10*, 1319 (1979); Anal. Abstr., *38*, 5D30 (1980)
35. Rao, P.D.: At. Absorpt. Newsl., *10*, 118 (1971)
36. Mulford, C.E.: At. Absorpt. Newsl., *5*, 88 (1966)
37. Barbooti, M.M., Jasim, F.: Talanta, *28*, 359 (1981)
38. Butler, L.R.P., Mathews, P.M.: Anal. Chim. Acta, *36*, 319 (1966)
39. Donaldson, E.M.: Talanta, *27*, 79 (1980)
40. Chau, Y.K., Lum-Shue-Chan, K.: Anal. Chim. Acta, *48*, 205 (1969)
41. Kim, C.H., Owens, C.M., Smythe, L.E.: Talanta, *21*, 445 (1974)
42. Kim, C.H., Alexander, P.W., Smythe, L.E.: Talanta, *23*, 229 (1976)
43. Kim, C.H., Alexander, P.W., Smythe, L.E.: Talanta, *22*, 739 (1975)
44. Yudelevich, I.G., Shaburova, V.P.: Chem. Anal. (Warsaw), *19*, 941 (1974); Chem. Abstr., *82*, 80043x (1975)
45. Schweizer, V.B.: At. Absorpt. Newsl., *14*, 137 (1975)
46. Sutcliffe, P.: Analyst (London), *101*, 949 (1976)
47. Caboi, R.: Period. Mineral., *37*, 717 (1968); Chem. Abstr., *71*, 27200t (1969)
48. Ise, K.: Bunseki Kagaku, *27*, 295 (1978)
49. Korkisch, J., Gross, H.: Talanta, *20*, 1153 (1973)
50. Pollock, E.N.: At. Absorpt. Newsl., *9*, 47 (1970)
51. Dreesen, D.R., Gladney, E.S., Owens, J.W.: J. Water Pollut. Control Fed., *51*, 2447 (1979)
52. Tse, K.: Bunseki Kagaku, *30*, 629 (1981)

53. Wada, K.: Bunseki Kagaku, *21*, 221 (1972)
54. Kirkbright, G.F., Smith, A.M., West, T.S.: Analyst (London), *91*, 700 (1966)
55. Endo, Y., Hata, T., Nakahara, Y.: Bunseki Kagaku, *18*, 878 (1969)
56. Nall, W.R., Brumhead, D., Whitham, R.: Analyst (London), *100*, 555 (1975)
57. Gregorczyk, S., Wycislik, A.: Chem. Anal. (Warsaw), *24*, 529 (1979); Chem. Abstr., *92*, 33254p (1980)
58. Thomerson, D.R., Price, W.J.: Analyst (London), *96*, 825 (1971)
59. Husler, J.: At. Absorpt. Newsl., *10*, 60 (1971)
60. Shevchuk, I.A., Tarsenko, L.E., Simonova, T.N., Naundorf, D.: Ukr. Khim. Zh., *43*(2), 198 (1977)
61. Li, S.-H., Wu, S.-C.: Fen Hsi Hua Hsueh, *8*, 438 (1980); Anal. Abstr., *42*, 5B140 (1982)
62. Castillo, J.R., Belarra, M.A., Aznarez, J.: At. Spectrosc., *3*, 58 (1982)
63. Welcher, G.G., Kriege, H.: At. Absorpt. Newsl., *8*, 97 (1969)
64. Saltykova, A.M., Davidovich, N.K., Melamed, Sh.G.: Zh. Anal. Khim., *27*, 1216 (1972)
65. Kudd, Y., Hasegawa, N., Yamashita, T.: Bunseki Kagaku, *20*, 1319 (1971)
66. Scarborough, J.M.: Anal. Chem., *41*, 250 (1969)
67. Pierce, J.O., Cholak, J.: Arch. Environ. Health, *13*, 208 (1966); Chem. Abstr., *65*, 12545b (1966)
68. Helsby, C.A.: Talanta, *20*, 779 (1973)
69. Roussos, G.G., Morrow, B.H.: Appl. Spectrosc., *22*, 769 (1968)
70. Henning, S., Jackson, T.L.: At. Absorpt. Newsl., *12*, 100 (1973)
71. Sang, S.L., Cheng, W.C., Shiue, H.I., Cheng, H.T.: Int. Sugar J., 77, 71 (1975); Chem. Abstr., *82*, 158097q (1975)
72. Neuman, D.R., Munshower, F.F.: Anal. Chim. Acta, *123*, 325 (1981)
73. Stupar, J., Dolinsek, F., Spenko, M., Furlan, J.: Landwirtsch. Forsch., *27*, 51 (1974); Chem. Abstr., *81*, 87422h (1974)
74. Stanton, R.E.: Lab. Pract., *23*(5), 233 (1974)
75. Bataglia, O.C.: Cienc. Cult. (Sao Paulo), *29*, 71 (1977); Chem. Abstr., *86*, 102709f (1977)
76. Daniel, R.C., Hänni, E., Shariatmadari, H.: Mitt. Geb. Lebensmittelunters. Hyg., *70*, 49 (1979); Anal. Abstr., *37*, 3G3 (1979)
77. Loon, J.C. van: At. Absorpt. Newsl., *11*, 60 (1972)
78. Fishman, M.J., Mallory, E.C.: J. Water Pollut. Control Fed., *40*, R 67 (1968)
79. Korkisch, J., Gödl, L., Gross, H.: Talanta, *22*, 669 (1975)
80. Korkisch, J., Krivanec, H.: Anal. Chim. Acta, *83*, 111 (1976)
81. Muzzarelli, R.A.A., Rocchetti, R.: Anal. Chim. Acta, *64*, 371 (1973)
82. Akama, Y., Nakal, T., Kawamura, F.: Nippon Kaisui Gakkaishi, *33*, 180 (1979), Chem. Abstr., *93*, 225380u (1980)
83. Nakachara, T., Chakrabarti, C.L.: Anal. Chim. Acta, *104*, 99 (1979)
84. Hrabovecka, G., Matherny, M.: Chem. Zvesti, *34*, 465 (1980)
85. Delaughter, B.: At. Absorpt. Newsl., *4*, 273 (1965)
86. Sterritt, R.M., Lester, J.N.: Analyst (London), *105*, 616 (1980)
87. Owens, J.W., Gladney, E.S.: At. Absorpt. Newsl., *15*, 95 (1976)
88. Hoover, W.L., Duren, S.C.: J. Assoc. Off. Anal. Chem., *50*, 1269 (1967)
89. Koirtyohann, S.R., Hamilton, H.: J. Assoc. Off. Anal. Chem., *54*, 787 (1971)
90. Leikovskaya, G.P., Danilova, V.D., Kutepova, A.I., Kon'kova, S.I., Yakovleva, P.F.: Khim. Prom-st., Ser.: Metody Anal. Kontrolya Kach. Prod. Khim. Prom-sti., (1), 34 (1980); Chem. Abstr., *93*, 106353c (1980)
91. Julietti, R.J., Wilkinson, J.A.: Analyst (London), *93*, 797 (1968)
92. Saba, C.S., Eisentraut, K.J.: Anal. Chem., *51*, 1927 (1979)
93. Kosonen, P.O., Salonen, A.M., Nieminen, A.L.: Finn. Chem. Lett., (4), 136 (1978)

Chapter 9

X-Ray Fluorescence

Present day x-ray spectrometers are compact, safe, reliable instruments. In some cases they are portable allowing one to make measurements at the sample site. Greatest analytical growth has been with x-ray fluorescence procedures. X-ray emission and absorption, although useful, are not as widely used as in the past. X-ray diffraction, essential for crystal structure measurements, is not generally applied to quantitative analysis. Auger spectrometry and other x-ray photoelectron techniques are not discussed here primarily because of limited access to these types of instruments by many analysts.

1. General Considerations

Listings of x-ray emission lines for molybdenum are available as part of exhaustive compilations of atomic lines [1–3]. Values for the more prominent of these are:

$K\alpha_1$	0.7093A	17.479 KeV
$K\alpha_2$	0.7136A	17.374 KeV
$K\beta_1$	0.6323A	19.608 KeV
$L\alpha_1$	5.4066A	2.293 KeV
$L\alpha_2$	5.4144A	2.290 KeV
$L\beta_1$	5.1771A	2.395 KeV

The $K\alpha_{1,2}$ line at 0.710 A is perhaps most often used for quantitative determination of molybdenum. Mass absorption coefficient for molybdenum $K\alpha_{1,2}$ is 18.1 cm^2/g and for $K\beta_1$ 13.3 cm^2/g [4].

Although emission measurement of molybdenum in ores [5] and absorption measurement in hydrocarbons [6] and aqueous solutions [7] are known, fluorescence measurements are probably used in the greatest number of x-ray procedures applied to molybdenum determination. Both wavelength dispersion and energy dispersion x-ray fluorescence lend themselves to molybdenum studies and x-ray tube and radioactive sources [8] and are used to excite the characteristic molybdenum emissions. Electron microprobe, too, in which secondary molybdenum emissions are initiated by bombardment from a highly focused beam of electrons, has been used for determining molybdenum [9–12].

Any x-ray fluorescence measurement is subject to interference from absorption and/or secondary emission of other constituents present in a sample [13]. Measurements upon the molybdenum peak (generally $K\alpha_{1,2}$) are no exception. Solid samples in particular must be carefully ground and polished if reproducible results are to be obtained [14, 15]. Mathematical corrections, taking into account a numerical correction term for each interfering element, serve to overcome matrix effects [16–19]. It is possible also to minimize matrix effects by observing secondary x-ray emissions from solutions rather than from solid samples [20]. Preconcentration of molybdenum with possible simultaneous removal of interfering ions by solvent extraction [21–23] and other separations also helps to avoid interelement interferences.

Standards for calibration of molybdenum fluorescence measurements should, as much as possible, mirror the unknown molybdenum containing sample. Not only should standards have the same elemental composition but components should be present in approximately the same proportions as in the unknown sample. Internal standards are also helpful to correct for instrument drift and other factors affecting the reproducibility of spectrometer readings.

2. Laboratory Procedures

Bustos has developed an x-ray fluorescent procedure for determining molybdenum in *copper molybdenum ore concentrates* and floation products [23]. To avoid matrix effects the molybdenum is extracted with an organic reagent. The reagent and recovery procedure are from a patented process [22] and utilize an α-hydroxy oxime as extracting agent. The reagent is 7-butyl-7-hydroxydodecan-6-one oxime (preparation given in the abstracting reference [22]) dissolved in tridecane. In selecting this reagent the author avoids the necessity of a second, back extraction of molybdenum into an aqueous phase or of having to destroy the organic material by ashing leaving an inorganic residue of molybdenum oxide which can be taken up in mineral acid solution. Measurements are made on the molybdenum $K\alpha_{1,2}$ emission peak at 0.710A. Background intensity on either side of the emission is used as an internal standard correction. Samples containing up to 160 fold excess copper and 80 fold excess iron over molybdenum do not interfere with the procedure. The calibration curve is linear from 5 to 700 mg Mo/L. Details of the procedure follow.

Operating Conditions

Instrumentation	Philips type PW 1212 spectrometer
X-Ray tube	chromium
	70 kv, 16 ma
Crystal	LiF (200)
Counter	scintillation
Measurement time	34 sec'

To prepare a calibration curve, dissolve 0.750 g chemically pure molybdenum oxide, MoO_3, in 10 mL of 0.1 M sodium hydroxide, NaOH. After dissolving, acidify the sample by adding 10 mL 9 M sulfuric acid, H_2SO_4. Dilute the sample to 100 mL in a volumetric flask. The stock solution contains 5.00 mg Mo/mL. A series of molybdenum standards is prepared by placing between 1.00 and 20.0 mL of stock solution into separate 100 mL volumetric flasks. These correspond to molybdenum concentrations between 5.00 and 100 mg Mo/100 mL. Add dropwise to each flask either 0.1 M H_2SO_4 or 0.1 M NaOH as appropriate until the pH of each solution is 1.45, read on a pH meter. Dilute each to exactly 100 mL with distilled water. Transfer 20.0 mL of each standard and a blank, containing only sulfuric acid adjusted to pH 1.45, to individual 125 mL separatory funnels. Add to each funnel 10 mL of 5% (w/v) oxime solution (in tridecane) and shake each funnel for two minutes. After the layers have separated, drain the organic phase into a 25 mL volumetric flask. Add a second 10 mL of 5% (w/v) oxime solution to each aqueous molybdenum sample and repeat the extractions. Combine the extracts in the volumetric flasks and add additional tridecane until each flask contain 25 mL of solution. Transfer a portion of each standard to a plastic sample cell with mylar (Dupont) window. Record the x-ray fluorescence intensity over the wavelength range centered at the $K\alpha_{1,2}$ Mo peak. The average background intensity (Record values on either side of the molybdenum peak, sum, and divide by two.), is substracted from the molybdenum intensity. Next the ratio of corrected molybdenum reading divided by the average background reading (as internal standard) is plotted vs. concentration of corresponding molybdenum standard.

A known portion of sample solution, obtained by appropriate sample treatment, is placed in a 100 mL volumetric flask. The pH of the sample is adjusted to 1.45 with either 0.1 M H_2SO_4 or 0.1 M NaOH as appropriate. The sample is diluted to exactly 100 mL and a 20.0 mL aliquot subjected to the same extraction procedure described for the standards. Concentration of molybdenum solution is read from the calibration plot and dilution corrections, if necessary, made to ascertain the percent molybdenum in the unknown sample.

To accurately account for matrix variations in the analysis of high temperature iron and nickel based alloys, Heinrich and Rasberry [17] employed a mathematical correction for interelement effects. Iron, nickel, chromium, cobalt, and molybdenum as major alloying elements are determined at levels that can exceed 10% by weight of the alloy composition. Ta, Al, Ti, Mn, Si, and V can be determined in the range 1 to 6% by weight. For molybdenum and the other major alloying elements listed, it is best if tungsten and/or niobium are absent from the sample. Measurements are made upon a series of molybdenum interfering element standards and from these correction coefficients determined which, when applied to the sample of interest, correct its intensity measurement for matrix effects. The procedure of Heinrich and Rasberry is not unique and other similar mathematic correction procedures are available [16]. Utilizing the H-R approach one finds for molybdenum in the presence of Ni, Co, Fe, Cr, and Ti.

$$\frac{C_{Mo}}{R_{Mo}} = 1 + \sum_{k=1}^{5} A_k C_k$$

where: C_{Mo} is the mass fraction of molybdenum present in the sample
C_k is the mass fraction of the kth interfering element
R_{Mo} is the measured relative x-ray intensity P, i.e.
$R_{Mo} = P_{Mo}/P_{Mo\ pure\ element}$
A_k is a coefficient for assessing the effect of the kth element upon the intensity of the molybdenum peak.

For molybdenum (one assumes $K\alpha_{1,2}$) the correction terms measured by Heinrich and Rasberry using various high temperature alloy standards are [17]:

element	Ni	Co	Fe	Cr	Ti
A_k	0.09	0.05	-0.02	-0.34	-0.84

The reader may wish to re-evaluate these constants using his/her own apparatus and/or particular alloys. The authors cite errors if 1 to 2% over a wide concentration range by this procedure. Following are details of the procedure.

Operating Conditions	
Instrumentation	Philips type 12045 spectrometer
X-Ray tube	tungsten
	45 kv, 10 ma
Crystal	LiF (200)
Counter	scintillation
Measurement time	100 sec

If correction coefficients for possible interfering elements are to be determined experimentally, binary standards of known composition for molybdenum with each interfering element are needed. In addition approximate composition of each interfering element in the unknown sample must be available to properly weight each correction term. A pure molybdenum metal sample is also necessary.

Samples and standards are in the form of highly polished disks approximately 3.1 cm in diameter and 2.5 cm high. Intensity readings are corrected for background counts before inserting them into the H-R formula. One solves the formula for C_{Mo}, the mass fraction of molybdenum present in the sample.

3. Applications to Specific Materials

Although not a trace technique for molybdenum in the μg/g or lower range, x-ray fluorescence measurements offer several advantages over other procedures for determining molybdenum. A comparison of a colorimetric and x-ray fluorescence method for molybdenum shows about equal overall accuracy but the fluorescence procedure required only one-tenth the amount of time [24]. X-ray measurements can be applied for a few percent to 100 percent of the sought for constituent. They are non-destructive. Solid samples remain essentially unchanged after exposure. With proper corrections for matrix effects, x-ray fluorescence yields accurate results. Table 9-1 lists various applications of x-ray fluorescence determination to specific materials.

Table 9-1. Determination of Molybdenum in Specific Materials by X-Ray Fluorescence Spectrometry

Material	Reference
rocks and minerals	
rocks	[25]
wolframite	[26]
Cu ore	[23, 27, 28]
ore	[29, 30, 31]
ore concentrate	[32]
corundum	[33]
ferrous alloys	
cast iron	[34]
ferromolybdenum	[35, 36]
Mo steel	[37]
Cr-Ni-Mo steel	[38]
stainless steel	[39]
low alloy steel	[40, 41]
high alloy steel	[42]
high temperature steel	[17]
tool steel	[43]
steel	[44, 45]
non ferrous alloys	
Ni alloy	[17]
Ta alloy	[46]
Ti alloy	[47]
U alloy	[48, 49]
W alloy	[50]
biological samples	
plants	[51]
environmental samples	
sea water	[52]
salt brines	[53]
miscellaneous samples	
hydrodesulfurisation catalyst	[54, 55, 56]
titanium carbide cermet	[57]
plutonium oxide cermet	[58]
mixed carbides Zr-Mo-U	[59]

References

1. Clark, G.L.: The Encyclopedia of X-Rays and Gamma Rays, 1963, New York, Reinhold Pub. Corp.
2. Bearden, J.A.: Rev. Mod. Phys., *39*, 78 (1967)
3. Blokin, M.A., Demekhin, V.F., Shveitser, I.G.: Izv. Akad. Nauk SSSR, Ser. Fiz., *28*, 834 (1964); Chem. Abstr., *61*, 12785b (1964)
4. Sweeney, W.R., Seal, R.T., Birks, L.S.: Spectrochim. Acta *17*, 364 (1961)
5. Fagel, J.E., Liebhafsky, H.A., Zemany, P.D.: Anal. Chem., *30*, 1918 (1958)

6. Barieau, P.E.: Anal. Chem., *29*, 348 (1957)
7. Bertin, E.P., Lingobucco, R.J., Carves, R.J.: Anal. Chem., *36*, 641 (1964)
8. Langheinrich, A.P., Foster, J.W.: Adv. X-Ray Anal., *11*, 275 (1967)
9. Poole, D.M.: Appl. Mater Res., *2*, 31 (1963)
10. Bastin, G.F., Heijwegen, C.P., Van Loo, F.J.J., Riech, G.D.: Mikrochim. Acta, 617 (1974)
11. Eberle, F., McCall, J.: Proc. Am. Power Conf., *26*, 488 (1964); Chem. Abstr., *62*, 8758f (1965)
12. Ilyin, N.P., Pozsgai, I.: Mikrochim. Acta Suppl., *8*, 213 (1979)
13. Kawashima, I.: Bunko Kenkyu, *17*, 22 (1968); Chem. Abstr., *70*, 43760k (1969)
14. Kopineck, H.J.: Arch. Eisenhüttenwes., *33*, 327 (1962); Chem. Abstr., *57*, 10852e (1962)
15. Tunney, A.A.: Br. Steel Corp. Open Rep., GS/EX/4/73/C (1973); Chem. Abstr., *81*, 57916e (1974)
16. Rasberry, S.D., Heinrich, K.F.J.: Anal. Chem., *46*, 81 (1974)
17. Heinrich, K.F.J., Rasberry, S.D.: Adv. X-Ray Anal., *17*, 309 (1974)
18. Mainardi, R.T., Fernandez, J., Bonetto, R., Riveros, J.A.: X-Ray Spectrom., *10*, 74 (1981)
19. Plesch, R.: X-Ray Spectrom., *10*, 193 (1981)
20. Pierron, E., Munch, R.: Develop. Appl. Spectrosc., *2*, 360 (1963)
21. Iwasaki, K., Tanaka, K., Takagi, N.: Bunseki Kagaku, *23*, 1179 (1974)
22. Swanson, R.R.: U.S. Patent 3,449,066 (1969); Chem. Abstr., *71*, P51733q (1969)
23. Bustos, L.I.: Analusis, *6*, 75 (1978)
24. Lingard, A.L., Willigman, M.G.: Proc. S. Dakota Acad. Sci., *42*, 170 (1963); Chem. Abstr., *60*, 11370a (1964)
25. Stern, W.B.: X-Ray Spectrom., *5*, 56 (1976)
26. Lipskaya, V.I., Ivoilov, A.S., Zav'yalova, L.L.: Nauch. Tr. Irkutsk. NII Redk. Tsvet. Met., (28), 71 (1976); Anal. Abstr., *33*, 6B174 (1977)
27. Smirnov, V.N., Ushkova, M.I., Novikov, A.M.: Zavod. Lab., *34*, 1326 (1968)
28. Yuan, H., Wu, C., Pu, S., Zhang, H., Xu, P., Wu, Q.: Fen Hsi Hua Hsueh, *9*, 146 (1981); Anal. Abstr., *41*, 4B34 (1981)
29. Gallagher, M.J.: Inst. Mining Met. Trans., Sec. B, *77*, 129 (1968); Chem. Abstr., *70*, 43772r (1969)
30. Bochenin, V.I.: Zavod. Lab., *42*, 433 (1976)
31. Bochenin, V.I.: Zavod. Lab., *44*(8), 960 (1978)
32. Studennikov, Yu.A., Belova, R.A.: Zavod. Lab., *33*, 1504 (1967)
33. Blank, A.B., Shevtsov, N.I., Mirenskaya, I.I.: Zh. Anal. Khim., *36*(7), 1272 (1981)
34. Kawashima, I., Tokiwa, K.: Nippon Kinzoku Gakkaishi, *29*, 1201 (1965); Chem. Abstr., *67*, 70260s (1967)
35. Bianchi, G., Grimaldi, R., Randi, G.: Metall. Ital., *69*, 435 (1977); Anal. Abstr., *35*, 1B152 (1978)
36. Staats, G.: Arch. Eisenhüttenwes., *45*, 693 (1974); Chem. Abstr., *82*, 50999w (1975)
37. Turner, P.J., Papazian, J.M.: Metal Sci. J., *7*, 81 (1973); Anal. Abstr. *26*, 1528 (1974)
38. Marti, W.: Spectrochim. Acta, *18*, 1499 (1963)
39. Momoki, K.: Bunseki Kagaku, *10*, 330 (1961)
40. Agrawal, R.M., Khanna, P.P.: Report BARC-699 (1973); Chem. Abstr., *82*, 92659r (1975)
41. Bowie, I.M., Brookes, L.: Br. Steel Corp. Open Rep., GS/TECH/403/1/73/C (1974); Chem. Abstr., *81*, 180895z (1974)
42. Luke, C.L.: Anal. Chem., *35*, 56 (1963)
43. Okochi, H., Takahashi, K., Suzuki, S., Sato, K., Sudo, E.: Trans. Natl. Res. Inst. Met. (Jpn.), *22*(11), 23 (1980); Chem. Abstr., *93*, 142238c (1980)
44. Dryer, H.T.: Adv. X-Ray Anal., *7*, 615 (1964)

45. Zitnansky, B.: Radioisotopy, *13*(6), 1135 (1972); Anal. Abstr., *26*, 3257 (1974)
46. Hakkila, E.A., Waterbury, G.R.: Talanta, *6*, 46 (1960)
47. Vasillaros, G.L., McKavaney, J.P.: Talanta, *16*, 195 (1969)
48. Karttunen, J.O.: Anal. Chem., *35*, 1044 (1963)
49. Florestan, J.: Method. Phys. Anal., 246 (1966)
50. Bayer, E.: Arch. Eisenhüttenwes., *47*, 157 (1976); Chem. Abstr., *85*, 71657k (1976)
51. Theisen, A.A., Pinkerton, A.: Soil Sci. Soc. Am. Proc., *32*, 440 (1968)
52. Monien, H., Bovenkerk, R., Kringe, K.P., Roth, D.: Fresineus' Z. Anal. Chem., *300*, 363 (1980)
53. Verbeeck, J., Vanderborght, B., Van Grieken, R., Ex, G.: Anal. Chim. Acta, *128*, 207 (1981)
54. Gunn, E.L.: J. Phys. Chem., *62*, 928 (1958)
55. LaBrecque, J.J.: J. Radioanal. Chem., *41*, 127 (1977)
56. LaBrecque, J.J., Pena, C.A., Marcano, E., Parker, W.C.: J. Radioanal. Chem., *60*, 247 (1980)
57. Wagner, J.C., Violante, E.J.: Appl. Spectrosc., *19*, 195 (1965)
58. Woltermann, H.A., Strohm, W.W.: Anal. Chem., *46*, 1822 (1974)
59. Hakkila, E.A., Hurley, R.G., Waterbury, G.R.: US At. Energy Comm., CONF-285-3 (1961); Chem. Abstr., *61*, 11332e (1964)

Chapter 10

Voltammetry

Voltammetry, in which the extent of a chemical change is monitored by the current observed as a function of an applied potential, has, as the result of advances in electronics, developed into sensitive and selective procedures for the study of reducible species. Classical direct current polarography with its linear ramp potential applied to a dropping mercury electrode is now supplemented with various other types of voltammetry. Alternating current polarography, differential pulse polarography, anodic stripping, and others are all used in the determination of molybdenum.

1. Polarography

Molybdenum(VI) exhibits two well defined polarographic waves at the dropping mercury electrode in solutions containing between 0.1 M and 2 M sulfuric acid representing reduction to Mo(V) and Mo(III) respectively [1]. $E_{1/2}$ values for these waves are at $+0.09$ v and -0.32 v vs. SCE in 1 M sulfuric acid. At other concentrations of H_2SO_4 or in the presence of additional sulfate salt either or both of these waves split and their $E_{1/2}$ values shift, probably because of complexation between various molybdenum species and sulfate [2–5]. In 1 M phosphoric acid two molybdenum waves with $E_{1/2}$ values of -0.33 v and -0.90 v vs. SCE occur [2]. The locations of these waves change with increasing phosphoric acid concentration [6]. In hydrochloric acid medium one or three waves are obtained [7, 8]. With 0.01 to 0.1 M HCl $E_{1/2}$ values of -0.10 v, -0.48 v, and -0.88 v vs. SCE appear for molybdenum within the concentration range (5 to 10) $\times 10^{-5}$ M [9]. As the HCl concentration increases only one wave persists with $E_{1/2} = -0.080 + 0.12 \log$ [H^+]$+0.035 \log$[Cl^-] volt vs. SCE in solutions 0.5 M to 0.7 M in HCl [10]. Molybdenum, first reduced to Mo(V) with hydrazine in 2 M HCl, exhibits further reduction polarographically with $E_{1/2}$ equal to -0.68 v vs. SCE [11].

In the presence of oxidizing agents the reduced forms of molybdenum, following polarographic reduction, are oxidized back to higher oxidation states only to be reduced again at the dropping mercury electrode. The result of this is enhanced limiting current values and, for analytical purposes, greater sensitivity for molybdenum. The limiting, maximum, current is now determined by the rate at which the oxidizing agent reacts with molybdenum, a catalytic process rather than a diffusion controlled process. The net change is the consumption of oxidizing substance even though molybdenum is reacting at the electrode surface. Several oxidizing agents cause well defined catalytic molybdenum waves. Of these perchlorate and nitrate are perhaps the best known. In 0.75 M H_2SO_4–1 M $HClO_4$ the observed wave with $E_{1/2}$ of -0.32 v vs. SCE is catalytically controlled [12]. In 0.1 M H_2SO_4–0.2 M Na_2SO_4–0.05 M HNO_3 a catalytic wave at -0.75 v vs. SCE occurs [13]. With 6 M HNO_3 only, the catalytic wave for molybdenum has its $E_{1/2}$ at -0.15 v vs. SCE [14]. Perchlorate [15] and nitrate [16] salts rather than $HClO_4$ or HNO_3, if in the presence of other acids, are equally satisfactory in oxidizing reduced forms of molybdenum. With perchlorate the catalytic wave results following oxidation of Mo(IV) to Mo(V) [15] but with nitrate present oxidation from Mo(III) to Mo(V) [15] and Mo(IV) to Mo(V) [17] are both reported. Bromate ion in 0.4 M acetic acid supporting electrolyte (E peak -0.44 v vs. SCE) [18], chloric acid (E peak -0.12 v vs. SCE) [19], and hydrogen peroxide in 0.05 M sulfuric acid (E peak $+0.28$ v vs. SCE) [20–22] also cause catalytic molybdenum waves.

Molybdenum heteropoly compounds are reducible at the dropping mercury electrode. Their use in determining both molybdenum and the hetero ion have recently been reviewed [23]. The molybdovanadophosphate complex, for example, in 1 M H_2SO_4 exhibits a reduction wave with $E_{1/2}$ -0.35 v vs. SCE which is used for determining molybdenum [24].

Complex ion formation prior to obtaining polarographic data results in a shift of the $E_{1/2}$ value to more negative potentials. Should two reducible ions in a non-complexing medium exhibit nearly identical $E_{1/2}$ values, it may be possible to determine one or both independently provided a suitable complexing agent can be found which sufficiently shifts the $E_{1/2}$ value of one of them. Various complexing agents have been added to molybdenum containing solutions in the hope of either shifting the $E_{1/2}$ value of molybdenum or of the interfering ion so molybdenum can be determined accurately. Frequently, for molybdenum, these complexing agents are hydroxy containing organic compounds. Table 10-1 lists several complexing agents used in determining molybdenum.

Besides complexation, other means of separating interfering elements prior to molybdenum measurement are used. Extracting agents for molybdenum are described in the following paragraph. Ion exchange separation prior to complexation with tartrate in the presence of sulfuric acid is successful for determining molybdenum in the presence of chromium [51]. Reduction of iron(III) to iron(II) with hydrazine followed by precipitation allows several metals, including molybdenum, to be measured polarographically [52]. Table 10-2 cites procedures for polarographic determination of molybdenum with other specific metals present in the sample matrix.

Polarographic measurement of molybdenum in mixed and nonaqueous solvents is possible. Generally a methanol solution of LiCl is used as supporting electrolyte. Molybdenum(V) thiocyanate complex is, for example, extracted with diethyl ether, 0.5 M LiOH (methanol) added, and a polarogram recorded using a DME-mercury pool electrode pair [65]. The procedure is suitable for steel samples. Water-acetonitrile and water-dimethyl sulfoxide yield polarographic waves for molybdenum [66]. Molybdenum complexes with catechol [29], 8-hydroxyquinoline [67], N-cinnamylphenylhydroxylamine [68], and N-benzoylphenylhydroxylamine [69, 70] are all extractable into organic solvents and then measured polarographically. Molybdenum 8-hydroxyquinoline is extractable in molten naphthalene. The solidified extract is dissolved in dimethylformamide and $HClO_4$-$NaClO_4$ added before polarography [71].

Table 10-1. Complex Forming Agents for Polarographic Determination for Molybdenum

Reagent	Condition	$E_{1/2}$ vs SCE, volts	Concentration Range, µg Mo/mL	Interference	Ref.
acetylacetone, 1 M	1 M H_2SO_4				[25]
aspartic acid, 0.1 M	pH 4, HCl	-0.62	1–100	Co, Cu, Ni	[26]
$NaBrO_3$, 0.25 M +2,4-dihydroxyaceto-phenone, 0.01 M	pH 5, acetate	-0.7^a	0.04–0.24	Cd, Pb	[27]
$KBrO_3$, 0.05 M +res-acetophenone isoniazid hydrazone, 0.0004 M MeOH	pH 5 acetate	-0.7^a	0.04–0.16	Fe, V	[28]
catechol, 0.2 M +H_2SO_4, 0.05 M	extract with 0.001 M $BuPh_3PBr$ ($C_2H_4Cl_2$), add 0.15 M diphenylquanidine +0.15 M HOAc (EtOH)	-0.42	0.4–40	Ti, V	[29]
$KClO_3$, 0.001 M +amygdalic acid, 0.002 M	0.5 M H_2SO_4	-0.62^b	0.001–		[30]
$KClO_3$, 0.25 M +glutaric acid, 0.02 M	0.1 M H_2SO_4 +0.2 M Na_2SO_4	-0.28^a	0.1		[31]
$KClO_3$, 0.1 M +glycollic acid, 0.1 M	0.04 M H_2SO_4 +0.2 M Na_2SO_4	-0.5^a	0.08–		[32]
$NaClO_3$, 0.2 M +hydrogen phthalate, K^+, saturated	pH 3.0	-0.6^a	20–800		[33]
$KClO_3$, 0.5 M +mandelic acid, 0.05 M	0.1 M H_2SO_4 +0.15 M Na_2SO_4	-0.26^b	0.001–	Cr, V	[34]
$KClO_3$, 0.25 M +trihydroxyglutaric acid, 0.075 M	0.1 M H_2SO_4 +0.2 M Na_2SO_4	-0.3^c	0.01–0.1	PO_4^{3-}, Sn	[35]
citric acid, 0.5 M	pH 1.8 HCl	-0.37^b	1–	Cu, Fe	[36, 37]
crotonic acid, 0.5 M	pH 3.5			Cd, Cr, In	[38]
3,4-dihydroxybenzene sulfonic acid, 0.1 M	pH 3.7, formate		5–20		[39]
ferron, 0.02 M	pH 4.0, acetate	-0.63	10–130	V, W	[40]
ferron, 0.02 M	pH 4.0 acetate	-0.63^c	0.01–0.1		[41]
glycollic acid, 0.1 M	pH 7, borate-phosphate	-1.2	10–1000		[42]
lactic acid, 0.5 M	pH 2.1	-0.22	20–	Cr, Fe, Se	[43]
malic acid, 0.05 M		-0.23	100–	W	[44]
mandelic acid, 0.1 M	0.1 M H_2SO_4 +0.15 M Na_2SO_4	-0.26^b	0.05–1	V, Ti, U	[45, 46]

Table 10-1. (continued)

Reagent	Condition	$E_{1/2}$ vs SCE, volts	Concentration Range, µg Mo/mL	Interference	Ref.
nitrilotriacetic acid, 0.15 M	pH 3 HCl	−0.48		Cr, Cu, Ni	[47]
NaIO₄, 0.2 M + thioacetamide, 0.002 M	pH 3.0, H₂SO₄	−0.88[a]	0.1–100		[48]
succinic acid, 0.1 M	pH 2.2	−0.51			[49]
tartaric acid, 0.1 M	pH 2.0	−0.52	20–350	NO₃⁻, Cr, Fe	[50]

[a] catalytic maximum
[b] alternating current polarography
[c] oscillographic polarography

Table 10-2. Polarographic Determination of Molybdenum in the Presence of Other Metals

Other Metal Present	Condition	Ref.
Cr	KNO₃	[53]
Nb	H₃PO₄	[54]
Sn	–	[55]
Ti	H₃PO₄, EtOH	[56]
Ti, Nb	H₂SO₄, citric acid, Th(NO₃)₄	[57]
Ti, W	H₃PO₄, HCl	[58]
W	pH 7, citric acid	[59]
W	pH 4.6, acetate	[60]
W	0.05 M H₂SO₄, mandelic acid	[61]
U	pH 1, HCl, NH₄Cl, H₂NNH₂	[62]
U	H₃PO₄	[63]
U	pH 4, acetate, ferron	[41]
V	H₂SO₄, Na₂SO₄	[4]
V	NaBrO₃, 3,4-dihydroxyacetophenone	[27]
V	H₂SO₄, resorcinol	[64]

Bhowal and Umland report a direct current polarographic procedure for molybdenum which, because of an initial extraction step, is moderately free from interferences [70]. The molybdenum sample is extracted by N-benzoylphenylhydroxylamine in chloroform. Aqueous perchloric acid and methanolic lithium chloride are added and a polarogram obtained. $E_{1/2}$ is −0.425 v vs. SCE. They successfully applied their procedure to a molybdenum containing steel. Ten replicate measurements upon the steel sample resulted in an average current of 0.743 µA with a standard deviation of 0.015 µA. A five fold or greater concentration of tungsten over molybdenum, in the presence of tartrate, interferes. Niobium, titanium, and rhenium in somewhat greater excess than tungsten also interfere. Ascorbic acid, added to mask iron(III), vanadium(V), and cerium(IV), results in a greater current for a given molybdenum concentration although the shape of the molybdenum wave is unchanged. The limiting diffusion current is linear with concentration over the range 0.5 to 10 µg Mo/mL. Following are details of their procedure.

Operating Conditions

Instrumentation	Tacussel model PRG5 polarograph
Electrodes	working electrode
	dropping mercury,
	m = 0.784 mg/sec
	t = 2.0 sec/drop (controlled)
	counter electrode
	Pt
	reference electrode
	saturated calomel
Temperature	23 °C
Deaeration	N_2 for 10 min prior to measurement

A *standard molybdenum solution* is prepared by dissolving 0.2 g ammonium molybdate, $(NH_4)_6Mo_7O_{24} \cdot 4H_2O$, in 500 mL of distilled water. A known aliquot of the solution is standardized by an appropriate procedure. Once the exact concentration is known the remaining stock solution is diluted with an appropriate amount of distilled water. The diluted solution should contain exactly 100 mg Mo (184 mg salt)/L. Between 0.10 mL (10 µg Mo) and 2.50 mL (250 µg Mo) of stock solution is placed into each of several 50 mL separatory funnels. One mL of 5 M hydrochloric acid, HCl, is added to each funnel and the contents of each funnel adjusted to 5 mL with distilled water. A blank solution containing 1 mL of 5 M HCl and 4 mL water is also prepared. 2.0 mL of 1% (w/v) N-benzoyl-N-phenylhydroxylamine, $C_6H_5CON(OH)C_6H_5$, in chloroform, $CHCl_3$, is added to each flask. Shake each flask 5 min. After the layers have separated, transfer each $CHCl_3$ layer to a separate 25 mL volumetric flask. Add 2.0 mL of pure $CHCl_3$ to each separatory funnel and again extract for 5 min. Combine each $CHCl_3$ layer with that from the original extraction in the 25 mL flask. To each flask add 5 mL 0.5 M perchloric acid, $HClO_4$, aqueous. 0.1 M lithium chloride, LiCl, in methanol is now added to each flask until the total volume of each is exactly 25 mL. Following deaeration with nitrogen, the polarogram of each standard and blank is recorded from −0.100 v to −0.800 v vs. SCE. From some convenient point on the limiting current plateau (e.g. −0.7 v vs. SCE) the current generated by each standard is recorded and plotted versus concentration of that standard.

The unknown sample is dissolved appropriately. A suitable aliquot is evaporated to near dryness and dissolved with 20 mL 3.5 M HCl. Filter if necessary and transfer the solution to a 100 mL volumetric flask. Dilute to 100 mL with distilled water. A 5.00 mL aliquot of this solution is extracted, etc. as described in the preceeding paragraph. From the calibration curve, the amount of molybdenum in the aliquot is determined. This value is corrected for the necessary dilutions and the amount of molybdenum in the unknown sample calculated. For samples containing large amounts of iron, 100 mg ascorbic acid is added to each sample and standard prior to obtaining the polarograms. In the presence of Nb, W, and/or Zr the N-benzoylphenylhydroxylamine is 2% (w/v) rather than 1%.

Differential pulse polarography, because of its increased sensitivity, is favored over conventional dc polarography for determining trace amounts of molybdenum. Lanza, Ferri, and Buldini have applied a differential pulse method for molybdenum in steels [72]. Their measurements are made upon the catalytic molybdenum wave observed in the presence of nitrate containing solutions. Fe(III), Mn(VII), Cr(VI), V(V), Cu(II), and Bi(III) do not interfere if they are present in amounts no greater than 100 times that of molybdenum. Steels containing more than 1% Mo can, for example, be analyzed without prior separation. For samples in which any of these interferences is present in amounts of greater than 100 times the molybdenum content, especially with iron, an ion exchange separation removes the interference before obtaining the polarogram. Tungsten in amounts 200 times that of molybdenum does not interfere providing no tungsten precipitate forms during sample dissolution. A precipitate would adsorb molybdenum causing low results. Also, if tungsten is present, the sample should be measured at once for upon standing or heating, the interfering effects of tungsten increase, probably because of some compound formation between the two species. Complexing agents that react with molybdenum cause a decrease in sensitivity. A standard addition technique is recommended by the authors to overcome any serious effects from their presence. Peak maximum for molybdenum in 2 M NH_4NO_3–0.25 M HNO_3 supporting electrolyte is -0.21 v vs. SCE. Detection limit is 5 ng/mL (5×10^{-8} M) molybdenum. Peak height is temperature sensitive with a temperature coefficient of about 7.5% per °C. A thermostated cell is recommended. Following are details of their procedure.

Operating Conditions

Instrumentation	AMEL model 472 polarograph
Electrodes	working electrode
	dropping mercury,
	m = 2.28 mg/sec
	t = 2.0 sec/drop (controlled)
	counter
	Pt ring (Ingold 805/NS)
	reference electrode
	saturated calomel
	(Ingold 303/NS)
Temperature	25.0±0.1 °C
Ramp potential	0 to -1.0 v vs. SCE, 2 mv/sec
Pulse potential	50 mv
Deaeration	N_2 for 5 min prior to measurement

A *standard molybdenum solution* is prepared by dissolving 0.4600 g ammonium molybdate, $(NH_4)_6Mo_7O_{24}\cdot4H_2O$, in doubly distilled water and diluting to 250 mL with additional water in a volumetric flask. This solution contains 1,000 µg Mo/mL. Transfer exactly 10.00, 1.00, and 0.10 mL of this stock solution to separate 100 mL volumetric flasks and dilute each to exactly 100 mL with distilled water. These diluted stock solutions contain 100, 10.0, and 1.0 µg Mo/mL respectively. Use whichever diluted standard solution that most closely matches the molybdenum content of the unknown molybdenum sample.

The unknown sample is dissolved by an appropriate means. If it contains no foreign ions in amounts sufficient to interfere with the molybdenum determination, one can omit the ion exchange separation. The sample should contain between 0.2 and 200 mg Mo. Evaporate the sample to near dryness. Add 10 mL 0.5 M nitric acid, HNO_3, and heat to dissolve the sample. Cool and transfer the sample to a 100 mL volumetric flask using 0.5 M HNO_3 to assist in the transfer. Add additional 0.5 M HNO_3 until exactly 100 mL of solution is obtained. Exactly one mL of this solution is transferred to a second 100 mL volumetric flask and again diluted to final volume with 0.5 M HNO_3. 50.0 mL of this is placed in a suitable vessel and 50.0 mL of 4 M ammonium nitrate, NH_4NO_3, solution added. The sample is deaerated for 5 min with nitrogen and the differential pulse polarogram recorded. To the remaining 50.0 mL aliquot add a known amount of one of the molybdenum standards. The standard chosen and the amount selected will depend upon the approximate molybdenum content of the unknown. Should, for example, the diluted unknown contain 10.0 µg Mo, 5.0 µg will remain in the unused aliquot. 5.00 mL of the 1.0 µg Mo/mL standard would then be appropriate. The amount of molybdenum added should approximate the amount present in the unknown aliquot. Dilute the sample plus standard to exactly 100 mL with 4 M NH_4NO_3 solution. Record the polarogram of the unknown plus standard sample.

If the unknown sample contains more than a 100 fold excess of iron or other interferring ion over molybdenum, evaporate the sample to near dryness and then add 20 mL 4 M HNO_3. Heat to dissolve the sample. The sample should contain between 0.1 and 100 mg Mo. Transfer the sample to a 50 mL volumetric flask and carefully add distilled water until the final volume is 50 mL. Prepare a strong cation resin column (Amberlite IR-120, 18–52 mesh) about 20 cm high in a 0.8 cm i.d. column. Slowly pass about 200 mL 0.5 M HNO_3 through the column to assure the resin is in the hydrogen form. Allow the liquid level above the resin to fall within one cm of the resin bed and then slowly add 10.0 mL of the molybdenum solution from the 50 mL volumetric flask. Elute the sample by slowly passing 0.5 M HNO_3 through the column collecting the eluent in a 100 mL volumetric flask. Continue passing 0.5 M HNO_3 through the column until the flask is full. Molybdenum is removed from the column while iron and other cations are retained. 50.0 mL of this solution is placed in a suitable vessel and 50.0 mL of 4 M NH_4NO_3 added. Deaeration and recording of the polarogram proceed as described in the proceeding paragraph. Molybdenum standard is added to the remaining 50.0 mL aliquot and a second polarogram recorded.

The height of the molybdenum peak, h_x, from the aliquot containing only the unknown portion of molybdenum, m_x, is proportional to the height of the molybdenum peak, h_{x+s}, from the aliquot containing both unknown and added standard molybdenum, $m_x + m_s$, where h represents peak height and m the concentration of molybdenum, µg/mL.

$$\frac{h_x}{h_{x+s}} = \frac{m_x}{m_x + m_s}.$$

From this, and considering the volume, V, of unknown and standard, one obtains:

$$m_x = \frac{h_x m_s V_s}{(h_{x+s}) V_t - h_x V_x}$$

where

m_x = concentration of unknown in 50 mL aliquot, µg/mL
m_s = concentration of standard, µg/mL
V_x = 50.0 mL
V_s = volume of standard added, mL
V_t = total volume, 100 mL
h_x = peak height unknown solution only
h_{x+s} = peak height unknown plus standard solution.

The value of m_x, once obtained, must be corrected further to account for the dilution steps prior to the final 100 mL dilution before the amount of molybdenum in the original unknown sample is found.

2. Anodic Stripping Voltammetry

Anodic stripping is a convenient and sensitive procedure for determining minute amounts of a reducible species. A hanging mercury drop electrode or other electrode type is adjusted to a potential sufficiently negative to deposit the metal of interest. Following deposition the potential of the electrode is reversed and a positive potential ramp, starting at the applied potential and proceeding fairly rapidly to more positive values, is begun. At its decomposition potential the deposited metal is re-oxidized and enters the solution. The name anodic stripping originates from the stripping of the deposited sample. Under these conditions the magnitude of the current observed, peak height, as the metal is stripped from the electrode, is proportional to its concentration. From a calibration curve of peak height versus known metal concentration, unknowns can be determined.

Several procedures for determining molybdenum by anodic stripping are available [73–75]. Of these the procedure of Ogura and Enaka [75] is described in detail. Their procedure uses a hanging mercury drop electrode for deposition of molybdenum. Following deposition the polarity of the electrode is reversed and a positive ramp potential applied. The stripping peak maximum is observed at a potential of approximately -0.45 v vs. SCE under the experimental conditions employed. Peak height is linear with molybdenum concentration when plotted on a log-log graph for molybdenum concentrations between 0.7 and 20 µg Mo/mL. No mention is made of the effects of interfering ions. Should other ions be present in the sample that might behave in a manner similar to molybdenum, additional studies will be necessary to learn if interference occurs. Following are details of their procedures.

Operating Conditions	
Instrumentation	three electrode potentiostat with variable scan rate, X-Y recorder

Electrodes	working electrode
	  hanging mercury drop
	counter electrode
	  Pt square
	reference electrode
	  saturated calomel

Temperature	25.0 ± 0.1 °C
Solution medium	potassium dihydrogen phosphate-sodium tetraborate buffer, pH 5.80
Deposition potential	-1.2 v vs. SCE
Deposition time	30 min
Anodic ramp potential	linear, 10 mV/min
Deaeration	N_2 for one hour prior to measurement

A *standard molybdenum solution* is prepared by dissolving 1.261 g sodium molybdate, $Na_2MoO_4 \cdot 2H_2O$, in distilled water and diluting to a final volume of 1,000 mL in a volumetric flask. 10.00 mL of this solution is placed in a 100 mL volumetric flask and distilled water added to reach the 100 mL mark. This solution contains 50.0 µg Mo/mL.

A standard series of known molybdenum solutions is prepared by placing between 1.00 mL (50.0 µg Mo) and 20.00 mL (1,000 µg Mo) in separate 50 mL volumetric flasks. Add approximately 75 mL buffer (containing 0.1 M potassium dihydrogen phosphate, KH_2PO_4, and 0.05 sodium tetraborate, $Na_2B_4O_7 \cdot 10H_2O$). With a pH meter adjust the pH of each sample to 5.80 by dropwise addition of either 0.1 M phosphoric acid, H_3PO_4, or 0.1 M sodium hydroxide, NaOH, as needed. Add distilled water until each flask contains exactly 100 mL of solution.

The hanging mercury drop electrode is obtained by placing one drop of mercury from a dropping mercury electrode, onto the flat surface of a platinum wire embedded in a glass support rod. The tip of the platinum prior to afixing the drop is plated with mercury by the following procedure. The polished wire tip is washed with distilled water and placed in a solution of saturated potassium chloride, KCl. A potential of 0.4v negative is applied between the platinum and a saturated calomel electrode also immersed in the KCl solution. The potential is maintained for 3 min. The electrode is now washed with distilled water and placed in a 0.25 M mercury(II) nitrate, $Hg(NO_3)_2$, solution. A potential of 0.5v negative is impressed between the platinum and a saturated calomel electrode also immersed in the solution. After three minutes the electrode is removed from the mercury solution, washed, and returned to the saturated KCl solution. A potential of 0.4v negative is again impressed upon the Pt electrode in the KCl solution. Following this treatment the Pt electrode is placed in the solution to be measured, the mercury drop attached, and the various external electrode connections completed. It is necessary to use a fresh drop of mercury for each run, i.e. with each change of sample.

Each standard sample is deaerated for one hour with nitrogen before proceeding and an atmosphere of nitrogen maintained over the sample during each run. Molybdenum is deposited upon the hmde at a potential of 1.2v negative for 30 min. Following this the linear ramp potential starting at $-1.2v$ and proceeding to more positive values is begun with a sweep rate of 10 mV/min. One can expect to observe the stripping peak at about $-0.45v$ vs. SCE. Peak height at the peak maximum vs. molybdenum concentration is plotted on log-log paper. The curve is linear over the range 7×10^{-6} M (0.7 µg/mL) to 2×10^{-4} M (20 µg/mL) molybdenum.

The unknown molybdenum solution containing from 0.05 to 1 mg Mo in about 20 mL of sample is placed in a 50 mL volumetric flask. Approximately 15 mL of phosphate-borate buffer is added and the pH adjusted to exactly 5.80 with the dropwise addition of either 0.1 M H_3PO_4 or 0.1 M NaOH. The sample is diluted to exactly 50 mL with distilled water, deaerated, and subjected to the deposition and stripping procedure as described previously for the molybdenum standards. The concentration of molybdenum in the unknown aliquot is read from the calibration curve. Further corrections for dilution may be necessary depending upon the initial sample treatment.

3. Applications to Specific Materials

Direct current polarography and the newer, more sensitive differential pulse and other voltammetric techniques are used in conjunction with diffusion controlled and catalytic polarographic waves to determine molybdenum in a variety of substances. Some of these are listed in Table 10-3.

Table 10-3. Determination of Molybdenum in Specific Materials by Voltammetry

Material	Reference	Material	Reference
rocks and minerals		Fe	[72]
silicate rocks	[76]	In	[96]
rocks	[12]	Si	[97]
oil shale	[77]	Zr	[98]
molybdenite	[22, 78, 79]	metal oxides	
ore	[51, 80, 81]	Th-U oxides	[47]
ferrous alloys		biological samples	
cast iron	[72]	human plasma	[99]
Cr steel	[70]	milk	[100]
Cr-Ni-Nb steel	[82]	clover	[101]
stainless steel	[83, 84]	grasses	[102]
Mn steel	[85]	potato	[103]
Ni steel	[86]	vegetables	[104]
low alloy steel	[65]	plants	[105]
high alloy steel	[87]	soils	[106, 107, 108, 109]
steel	[24, 25, 38, 88]	environmental samples	
non ferrous alloys		natural water	[110]
Nb alloys	[89]	sea water	[111]
Ni alloys	[90, 91]	salt brines	[33]
Ti alloys	[92]	fly ash	[112]
U alloys	[93]	miscellaneous samples	
U-Nb-Zr alloys	[94]	drugs	[113]
Nb-Mo-Zr alloys	[57]	fertilizer	[114, 115]
Al-Mo-W-Ti alloys	[58]	glass	[116]
pure elements		glass ceramic	[117]
Cd	[95]	sulfuric acid	[118]

References

1. Kolthoff, I.M., Hodara, I.: J. Electroanal. Chem., *4*, 369 (1962)
2. Höltje, R., Geyer, R.: Z. Anorg. Allg. Chem., *246*, 258 (1941)
3. Li, P.C., Chao, T.H., Li, Y.H.: Chi-Lin Ta Hsueh Tzu Jan K'o Hsueh Hsueh Pao, (1), 99 (1959), Chem. Abstr., *56*, 10903i (1962)
4. Gupta, C.M., Gupta, J.K.: J. Indin Chem. Soc., *44*, 526 (1967)
5. Chavdarova, R.D.: Dokl. Bolg. Adad. Nauk, *32*, 1093 (1979); Chem. Abstr., *92*, 87524v (1980)
6. Kurbatov, D.I., Dieva, E.N.: Tr. Inst. Khim., Akad. Nauk SSSR, Ural. Fil., (17), 172 (1970); Chem. Abstr., *73*, 136638d (1970)
7. El-Shamy, H.K., Barakat, M.F.: Egypt. J. Chem., *2*, 101 (1959)
8. Wolter, M., Wolf, D., von Stackelberg, M.: J. Electroanal. Chem., *22*, 221 (1969)
9. Bikbulatova, R.U., Sinyakova, S.I.: Izv. Akad. Nauk Tadzh. SSR, Otd. Fiz. Mat. Geol. Khim. Nauk, (2), 48 (1969); Chem. Abstr., *73*, 51608u (1973)
10. Haight, G.P.: J. Inorg. Nucl. Chem., *24*, 673 (1963)
11. Ivanova, Z.I., Chernikova, E.N., Lektorskaya, N.A.: Zh. Anal. Khim., *28*, 2202 (1973)
12. Sinyakova, S.I., Glinkina, M.I.: Zh. Anal. Khim., *11*, 544 (1956)

13. Johnson, M.G., Robinson, R.J.: Anal. Chem., *24*, 366 (1952)
14. Edmonds, T.E.: Anal. Chim. Acta, *108*, 155 (1979)
15. Stach, B., Schöne, K.: Mikrochim. Acta, II, 565 (1977)
16. Christian, G.D., Vandenbalch, J.L., Patriarche, G.J.: Anal. Chim. Acta, *108*, 149 (1979)
17. Edmonds, T.E.: Anal. Chim. Acta, *116*, 323 (1980)
18. Toropova, V.F., Vekslina, V.A., Chovnyk, N.G.: Zh. Anal. Khim., *28*, 967 (1973)
19. Kolthoff, I.M., Hodara, I.: J. Electroanal. Chem., *5*, 165 (1963)
20. Kolthoff, I.M., Parry, E.P.: J. Am. Chem. Soc., *73*, 5315 (1951)
21. Lamache-Duhameaux, M.: C.R. Acad. Sci., Ser. C, *270*, 1193 (1970)
22. Sharipova, N.S., Songina, O.A.: Zh. Anal. Khim., *28*, 2348 (1973)
23. Alimarin, I.P., Dorokhova, E.N., Kazanskii, L.P., Prokhorova, G.V.: Zh. Anal. Khim., *35*, 2000 (1980)
24. Furtuné, L.A., Nguyen Van Chu, Polotebnova, N.A.: Zh. Anal. Khim., *29*, 1118 (1974)
25. Pantani, F.: Ric. Sci., *1*, 12 (1961); Chem. Abstr., *57*, 9203h (1962)
26. Gupta, C.M., Gupta, J.K.: Talanta, *15*, 274 (1968)
27. Rao, V.S.N., Rao, S.B.: Talanta, *26*, 502 (1979)
28. Mohan, K.M., Rao, S.B.: Fresenius' Z. Anal. Chem., *303*, 121 (1980)
29. Lenda, K., Bartusek, M.: Scr. Fac. Sci. Natur. Univ. Purkynianae Brun., *3*(5–6), 49 (1973); Chem. Abstr., *81*, 72210p (1974)
30. Zaitsev, P.M., Zaitseva, Z.V., Zhdanov, S.I., Nikolzeva, T.D.: Zh. Anal. Khim., *35*, 1951 (1980)
31. Chikryzova, E.G., Kiriyak, L.G.: Zh. Anal. Khim., *29*, 899 (1974)
32. Chikryzova, E.G., Bardina, S.M.: Zh. Anal. Khim., *29*, 2414 (1974)
33. Kyriacon, D.: Anal. Chem., *41*, 844 (1969)
34. Pakhomova, K.S., Volkova, L.P.: Zh. Anal. Khim., *31*, 952 (1976)
35. Chikryzova, E.G., Kiriyak, L.G.: Zh. Anal. Khim., *27*, 1747 (1972)
36. Vasil'eva, L.N., Pozdnyakova, A.A.: Sb. Nauchn. Tr. Gos. Nauchno-Issled. Inst. Tsvetm. Met., (27), 44 (1967); Chem. Abstr., *70*, 25430y (1969)
37. Moosmüller, A., Hohn, H.: Z. Anorg. Allg. Chem., *373*, 148 (1970)
38. Rao, A.L.J., Singh, M.: Trans. SAEST, *13*, 61 (1978); Chem. Abstr., *90*, 179527k (1979)
39. Zelinka, J., Bartušek, M., Okáč, A.: Collect. Czech. Chem. Commun., *38*, 2898 (1973)
40. Bertoglio Riolo, C., Fulle Soldi, T., Occhipinti, C.: Ann. Chim. (Rome), *57*, 1344 (1967)
41. Bertoglio Riolo, C., Fulle Soldi, T., Spini, G.: Ann. Chim. (Rome), *58*, 3 (1968)
42. Ogura, K., Enaka, Y., Morimoto, K.: Electrochim. Acta, *23*, 289 (1978)
43. Mahanti, H.S., Deshmukh, G.S.: Chem. Ind. (London), (1), 40 (1975)
44. Gupta, J.K., Gupta, C.M.: Inorg. Chim. Acta, *3*, 358 (1969)
45. Pakhomova, K.S., Volkova, L.P.: Zh. Anal. Khim., *31*, 329 (1976)
46. Mittal, M.L.: Z. Naturforsch. Teil B, *24*, 1053 (1969)
47. Manning, D.L., Ball, R.G., Menis, O.: Anal. Chem., *32*, 1247 (1960)
48. Farr, J.P.G., Laditan, G.O.A.: J. Less Common Metals, *36*, 161 (1974)
49. Gupta, C.M.: Bull. Chem. Soc. Jpn., *40*, 221 (1967)
50. Parry, E.P., Yakubik, M.B.: Anal. Chem., *26*, 1294 (1954)
51. Holzapfel, H., Gürtler, O., Tempel, B.: Fresenius' Z. Anal. Chem., *235*, 413 (1968)
52. Geyer, R.: Z. Anorg. Allg. Chem., *271*, 93 (1952)
53. Geyer, R., Henze, G., Henze, J., Müller, E.G.: Proc. Conf. Appl. Phys. Chem. Methods Chem. Anal., Budapest, *1*, 211 (1966); Chem. Abstr., *69*, 8221t (1968)
54. Kurbatov, D.I., Voronova, E.M.: Tr. Inst. Khim. Akad. Nauk SSSR, Ural. Fil., (10), 57 (1966); Chem. Abstr., *66*, 43443b (1967)

55. Bardina, S.M., Chikryzova, E.G., Radautsan, S.I., Samus, D.P.: Polyarogr. Tekh. Anal. Kishinev, 24 (1979); Anal. Abstr., *39*, 6B116 (1980)
56. Kurbatov, D.I., Nikitina, G.A.: Zh. Anal. Khim., *36*(4), 687 (1981)
57. Headridge, J.B., Hubbard, D.P.: Anal. Chim. Acta, *35*, 85 (1966)
58. Kurbatov, D.I., Il'kova, S.B., Tugusheva, G.A.: Zh. Anal. Khim., *32*, 741 (1977)
59. Meites, L.: Anal. Chem., *25*, 1752 (1953)
60. Monien, H., Jacob, P., Jänisch, B.: Fresenius' Z. Anal. Chem., *267*, 108 (1973)
61. Henrion, G., Scholz, F., Schmidt, R., Febian, I.: Z. Chem., *21*(3), 104 (1981)
62. Deshmukh, G.S., Srivastava, J.P.: Fresenius' Z. Anal. Chem., *176*, 28 (1960)
63. Kurbatov, D.I., Niktina, G.A.: Zh. Anal. Khim., *32*, 2082 (1977)
64. Etaiw, S.H., Issa, I.M., Ismail, M.I.: Ann. Chim. (Rome), *69*, 219 (1979)
65. Afghan, B.K., Dagnall, R.M.: Talanta, *14*, 239 (1967)
66. Sagradyan, S.I., Agasyan, P.K., Nikolaeva, E.R.: Vestn. Mosk. Univ. Khim., *16*(1), 67 (1975); Chem. Abstr., *83*, 172090j (1975)
67. Pyatnitskii, I.V., Ruzhanskaya. R.P.: Zh. Anal. Khim., *25*, 1063 (1970)
68. Shpak, E.A., Samchuk, A.I., Pilipenko, A.T.: Zh. Anal. Khim., *29*, 938 (1974)
69. Pilipenko, A.T., Shpak, E.A., Samchuk, A.I.: Zh. Anal. Khim., *30*, 1086 (1975)
70. Bhowal, S.K., Umland, F.: Fresenius' Z. Anal. Chem., *282*, 197 (1976)
71. Fujinaga, T., Puri, B.K.: Fresenius' Z. Anal. Chem., *269*, 340 (1974)
72. Lanza, P., Ferri, D., Buldini, P.L.: Analyst (London), *105*, 379 (1980)
73. Pnev, V.V., Popov, G.N., Nagarev, V.G.: Zh. Anal. Khim., *28*, 2050 (1973)
74. Bramin, V.A., Kaplin, A.A.: Izv. Tomsk. Politekhn. Inst., (275), 119 (1976); Anal. Abstr., *33*, B117 (1977)
75. Ogura, K., Enaka, Y.: J. Electroanal. Chem., *95*, 169 (1979)
76. Belova, T.Ya., Volkova, L.P., Pakhomova, K.S.: Zavod. Lab., *44*, 1176 (1978)
77. Bastova, S.M., Yurina, R.D., Vakhobova, R.U.: Zh. Anal. Khim., *34*, 935 (1979)
78. Galli, Z.A., Sheina, N.M., Polikarpova, N.V., Pleshakova, T.V.: Zh. Anal. Khim., *30*, 1148 (1975)
79. Holten, C.H.: Acta Chem. Scand., *15*, 943 (1961)
80. Yokosuka, S.: Bunseki Kagaku, *2*, 319 (1953)
81. Chen, Y.: Fen Hsi Hua Hsueh, *9*, 87 (1981); Anal. Abstr., *41*, 3B191 (1981)
82. Haight, G.P.: Anal. Chem., *23*, 1505 (1951)
83. Kurobe, M., Terada, H., Tajima, N.: Bunseki Kagaku, *11*, 767 (1962)
84. Sen, P.K.: Sci. Cult., *43*, 530 (1977)
85. Patriarche, G., Gerbaux, A.A., Molle, L.: Bull. Soc. Chim. Belges, *74*, 240 (1965)
86. Bhowal, S.K., Bhattacharyya, M.: Fresenius' Z. Anal. Chem., *310*, 124 (1982)
87. Pecsok, R.L., Parkhurst, R.M.: Anal. Chem., *27*, 1920 (1955)
88. Sverak, J.: Fresenius' Z. Anal. Chem., *201*, 9 (1964)
89. Headridge, J.B., Hubbard, D.P.: Analyst (London), *90*, 173 (1965)
90. Wolfson, H.: Nature, *153*, 375 (1944)
91. Kawahata, M., Mochizuki, H., Kajiyama, R., Irikura, K.: Bunseki Kagaku, *11*, 317 (1962)
92. Codell, M., Mikula, J.J., Norwitz, G.: Anal. Chem., *25*, 1441 (1953)
93. Athavale, V.T., Kalyanaraman, R., Khasgiwale, K.A.: Anal. Chim. Acta, *29*, 280 (1962)
94. Arragon, Y.: RAPP. CEA, R-4475 (1973); Anal. Abstr., *27*, 2510 (1974)
95. Lysenko, V.I.: Metody Anal. Vesh. Vysokoi Chistoty, Akad. Nauk SSSR, Inst. Geokhim. Anal. Khim., 382 (1965); Chem. Abstr., *65*, 4642c (1966)
96. Bikbulatova, R.U., Sinyakova, S.I.: Zh. Anal. Khim., *19*, 1434 (1964)
97. Buldini, P.L., Ferri, D.: Anal. Chim. Acta, *124*, 233 (1981)
98. Wood, D.F., Clark, R.T.: Analyst (London), *87*, 342 (1962)
99. Christian, G.A., Patriache, G.J.: Analyst (London), *104*, 680 (1979)

100. Gal'tseva, V.P.: Izv. Timiryazevsk. S'kh. Akad. No. 2, 184 (1968); Chem. Abstr., *68*, 18068n (1968)
101. Lapitskaya, S.K., Sviridenko, V.G.: Zh. Anal. Khim., *33*, 1583 (1978)
102. Nangniot, P.: J. Electroanal. Chem., *7*, 50 (1964)
103. Chatelet-Cuzin, A.M., Ecrement, F., Durand, G.: Analusis, *9*, 433 (1981)
104. Fichera, P., Ferrara, S.: Agrochimica, *13*, 85 (1969); Chem. Abstr., *71*, 100454k (1969)
105. Jones, G.B.: Anal. Chim. Acta, *10*, 584 (1954)
106. Samokhvalov, S.G.: Agrokhimiya, (5), 118 (1965); Chem. Abstr., *63*, 12263b (1965)
107. Bikbulatova, R.U., Sinyakova, S.I.: Agrokhimiya, (6), 123 (1967); Chem. Abstr., *67*, 97535x (1967)
108. Kocurik, S.: Agrochemia (Bratislava), *5*, 111 (1965); Chem. Abstr., *67*, 63313h (1967)
109. Ruchko, G.V., Skobets, E.M.: Khim. Sel'sk. Khoz, *4*, 14 (1966); Chem. Abstr., *64*, 16617h (1966)
110. Kuan, I.-W., Li, N.-C.: Hua Hsueh Tung Pao, (1), 15 (1974); Chem. Abstr., *82*, 7527d (1975)
111. Prabhu, V.G., Zaraphar, L.R., Das, M.S.: Mikrochim. Acta, II, 67 (1980)
112. Bosserman, P., Sawyer, D.T., Page, A.L.: Anal. Chem., *50*, 1300 (1978)
113. Molle, L., Patriarche, G., Gerbaux, A.A.: J. Pharm. Belg., *20*, 263 (1965); Chem. Abstr., *64*, 7972d (1966)
114. Protsenko, G.P., Kovalenko, P.N.: Agrokhimiya, (3), 127 (1966); Chem. Abstr., *64*, 20580b (1966)
115. Nabieva, M.M., Khakimova, V.K.: Uzb. Khim. Zh., *17*(1), 13 (1973); Chem. Abstr., *79*, 4190r (1973)
116. Andrusieczko, B.: Szklo Ceram., *29*, 143 (1978); Chem. Abstr., *90*, 59773r (1979)
117. Beleuta, I.L., Bratu, C.: Radiochem. Radioanal. Lett., *2*, 233 (1969)
118. Zaitsev, P.M., Dergacheva, E.N.: Khim. Prom-st., Ser.: Metody Anal. Kontrolya Kach. Prod. Khim. Prom-sti., (4), 38 (1980); Chem. Abstr., *93*, 106430a (1980)

Chapter 11

Catalytic Methods

1. Hydrogen Peroxide – Iodide Reaction

It is well known that molybdenum and other ions catalyze the reaction of hydrogen peroxide with iodide ion.

$$H_2O_2 + 2\,I^- + 2\,H^+ \rightarrow I_2 + 2\,H_2O$$

The formation rate of iodine increases uniformly with increasing presence of molybdenum for molybdenum concentrations between about 0.02 to 0.09 µg Mo/mL [1]. Iodine formation is followed by monitoring the increasing color intensity of the blue iodine-starch complex at 590 nm or of the triiodide ion, I_3^-, at 350 nm. Since first reported for determination of molybdenum [1], this application of the hydrogen peroxide-iodide reaction has been extensively studied. It is found, for example, that improved accuracy and precision are achieved if one pays close attention to experimental details. Thermostated baths, preliminary reagent purification, and carefully controlled reagent concentrations all contribute to the success of this method. Using mathematical optimization techniques, concentrations of reagents involved in the determination (H_2O_2, H_2SO_4, KI, starch) were adjusted for maximum desirable results [2]. Doing so resulted in an acceptable lower detection limit of 0.006 µg Mo/mL rather than the 0.02 µg Mo/mL originally reported. Values for reagent concentrations are compared in Table 11-1.

Traditionally the increasing absorbance of the iodine-starch complex is recorded as a function of time for several solutions each containing a known, fixed amount of molybdenum. All reagents except hydrogen peroxide are mixed together, allowed to reach thermal equilibrium, and timing started the instant the H_2O_2 solution is added. Absorbance values are conveniently read at one minute intervals. The absorbance versus time data for each sample are plotted. At some fixed time common to all curves and in the linear region of each curve, a tangent is constructed. Tangent values increase with increasing amounts of catalyst, that is to say the slope of the line representing the reaction rate increases with increasing catalyst concentration. A second plot of tangent value versus catalyst concentration serves as the calibration curve. The tangent value for an unknown molybdenum containing solution is converted to its corresponding molybdenum value from this graph.

Variamine blue is sometimes used in place of starch to form a colored complex with iodine [3]. This more intensely colored species has its absorption maximum at 550 nm. Amperometric monitoring of the I^-/I_2 reaction, in which current rather

Table 11-1. Reagent Conditions for Catalytic Determination of Molybdenum by the Hydrogen Peroxide-Iodide Reaction

	Original Values [1]	Optimum Values [2]
H_2O_2	0.001 M	0.00088 M
H_2SO_2	0.08 M	0.03 M
KI	0.02 M	0.0028 M
starch	0.02% (w/v)	0.02% (w/v)

than absorbance is related to the rate of iodine formation, frequently substitutes for colorimetric measurement [4–7]. Other means of following iodide/iodine conversion include thermometric [8] and potentiometric [9] measurements.

Because of the time and effort required to collect data and construct graphs, several modifications of the traditional catalytic procedure are possible. A single absorbance measurement at a predetermined time interval for each molybdenum catalyzed reaction substitutes for the tangent value and is plotted as a single calibration curve of absorbance value versus molybdenum concentration. An 18 min delay is recommended for the hydrogen peroxide-iodide reaction [10]. One must realize, however, that the reliability of a single measurement is much less than that obtained for a series of measurements. A better approach is to use what is known as a Landolt reaction [11]. Here a small amount of additional reagent is added which consumes the product monitored in the catalytic reaction. With molybdenum catalyzed oxidation of iodide by hydrogen peroxide, a constant amount of ascorbic acid is added to each sample used in generating the standard series of known molybdenum solutions and to the unknown molybdenum containing solution. Iodine, as soon as it forms, is reduced by ascorbic acid and the iodine-starch complex does not appear until all the ascorbic acid is used up. One measures the time from start of the reaction until the blue color appears. Because iodine is generated at a faster rate when greater amounts of molybdenum are present, this time interval will decrease; that is, the ascorbic acid will be consumed more quickly as the amount of molybdenum present increases. This modification is applied for molybdenum [12, 13]. Various other modifications [14, 15] and automated procedures [16–18] are also applied to molybdenum determination.

Iron(III) vanadium, titanium, tungsten, and chromium all to varying degrees catalyze the hydrogen peroxide-iodide reaction. Other metals which react with iodide and/or hydrogen peroxide must also be absent as must any oxidizing agent that converts I^- to I_2. Addition of fluoride reduces to some extent the interference of Fe, V, Ti, W, and Cr [19, 20]. Citric acid eliminates tungsten interference completely making possible the determination of molybdenum in the presence of tungsten [21]. Other catalytic procedures using reactions of hydrogen peroxide and other reducing agents also provide for the determination of molybdenum in the presence of tungsten [22, 23].

The following procedure is that of Altinata and Pekin for *catalytic determination* of molybdenum [9]. The course of the reaction is monitored by following the decrease in iodide concentration with an iodide specific ion electrode.

Reagents

0.050 M H_2O_2: Dilute 58 mL of redistilled 30% hydrogen peroxide, H_2O_2, with redistilled, deionized water to approximately 90 mL total volume. Standardize the sample with standard 0.050 M Ce(IV) solution. Based upon the results add additional water to prepare exactly 0.050 M H_2O_2 solution. The solution should be standardized every few days.

0.010 M KI: Potassium iodide, KI, is recrystallized twice from ethanol and dried at 110 °C. Exactly 0.166 g of KI is dissolved in redistilled, deionized water and diluted to a final volume of exactly one liter. The solution is 0.010 M in KI.

0.80 M H_2SO_4: Dilute 4.6 mL of 18 M reagent grade sulfuric acid, H_2SO_4, to approximately 90 mL with redistilled, deionized water. Standardize the solution with standard 0.080 M base and add additional water as needed to prepare exactly 0.080 M H_2SO_4.

1.0 µg/mL Mo solution: Dissolve in redistilled, deionized water 1.847 g ammonium paramolybdate, $(NH_4)_6Mo_7O_{24}·4H_2O$ and dilute to one liter in a volumetric flask. 1.00 mL of this solution is again diluted to one liter in a second volumetric flask. The diluted solution contains 1.0 µg Mo/mL.

Procedure

5.00 mL of 0.80 M H_2SO_4, 1.00 mL of 0.10 M KI, and sufficient redistilled, deionized water to achieve a final volume of about 45 mL are added to each of several beakers in a 25 ± 0.02 °C water bath. The beakers are calibrated to indicate exactly 50 mL of solution and are of sufficient size and shape to accommodate an iodide specific ion electrode and a reference electrode. With a micropipet, add varying but known amounts (between 0.25 and 1.00 mL) of molybdenum standard solution (containing 1.0 µg Mo/mL) to all but two of the beakers. One beaker should have no added molybdenum, the blank, and one beaker a specified amount of unknown molybdenum containing solution. Ideally the unknown is previously diluted to contain between 0.25 and 1.0 µg Mo/mL thus allowing the sample data to fall within that of the molybdenum standards. Dilute all samples to exactly 50 mL with water and insert the electrodes.

Allow each sample to remain for 10 min in the water bath before proceeding. Now add rapidly to each sample 1.00 mL of 0.050 M H_2O_2 solution. Begin timing each sample at the moment of contact between sample and peroxide solution. Follow the decrease in iodide concentration of each sample, as registered on a specific ion meter either as potential or as concentration reading, at appropriate time intervals until no further significant change in iodide level is observed. Plot meter reading versus time for each sample. From an appropriate time (identical for each sample) on the linear portion of each plot construct a tangent. Prepare a second plot of tangent value versus molybdenum concentration for each of the standards and blank. From the tangent value of the unknown molybdenum containing solution and the calibration plot, determine the concentration of unknown molybdenum. Correct this value for dilution from the original unknown sample and report amount of molybdenum present.

2. Other Molybdenum Catalyzed Reactions

Molybdenum catalyzes a variety of other reactions between hydrogen peroxide and various reducing agents. Some of these are detailed in Table 11-2.

Oxidation of various substances by bromate ion, BrO_3^-, is catalyzed by molybdenum and other ions. These reactions, too, find application in the determination of molybdenum [29]. Variation in the rate of the bromate-iodide reaction has, for example, been used in the analysis of mixtures containing Mo(VI) and Cr(VI) or W(VI) [30]. Acid phosphatase catalysis of naphthyl-1-phosphate hydrolysis to

Table 11-2. Catalytic Determination of Molybdenum by Its Effect Upon the Reaction of Hydrogen Peroxide and Various Reducing Agents

Reagent	Condition	Indication	Ref.
2-aminophenol	pH 5.5–6.0, acetate	photometric, $\lambda = 430$ nm	[23]
azo rubin S	pH 10.3, carbonate	photometric, $\lambda = 540$ nm	[24]
iodide	0.08 M H_2SO_4	photometric, $\lambda = 590$ nm	[1, 25]
rubeanic acid	0.1 M HCl	photometric, $\lambda = 360$ nm	[26, 27]
thiosulfate	pH 3.4, acetate	thermometric	[22]
		potentiometric	[28]

Table 11-3. Catalytic Determination of Molybdenum by Its Effect Upon Various Reactions

Reaction	Condition	Indication	Ref.
bromate + 1-amino-2-naphthol-4-sulfonic acid	pH 1.85	photometric, $\lambda = 250$ nm	[32]
bromate + iodide	neutral	amperometric, 2 Pt electrodes, 50 mV	[33, 34]
bromate + naphthylamine	pH 2, H_2SO_4	photometric, $\lambda = 530$ nm	[35, 36]
malachite green + Ti(III)	pH 1.7, H_2SO_4	photometric, $\lambda = 617$ nm	[37, 38]
methylene blue + hydrazine	0.36 M H_2SO_4	photometric, $\lambda = 670$ nm	[39, 40]
perborate + iodide	pH 4.0, acetate	visual, I_2-starch	[41]
peroxyacetic acid + ascorbic acid	pH 4.2 iodide	potentiometric	[42]
selenate + Sn(II)	1 M HCl	turbidimetric, $\lambda = 390$ nm	[43]

Table 11-4. Catalytic Determination of Molybdenum in Specific Materials

Material	Reaction	Ref.
rocks and minerals		
rocks	$H_2O_2 + I^-$	[44]
ferrous alloys		
steel	$H_2O_2 +$ rubeanic acid	[26]
biological samples		
blood	$SeO_4^{2-} + Sn(II)$	[45]
cotton	inhibit Re catalysis of α-furildioxime hydrolysis	[46]
tomato leaf	$H_2O_2 + I^-$	[19]
rye grass	$H_2O_2 + I^-$	[20]
environmental		
sea water	$H_2O_2 + I^-$	[47]
sea water	$H_2O_2 +$ rubeanic acid	[27]
sea water	$BO_3^- + I^-$	[48]
sea water	Fe(III) tartrate complex + Sn(II)	[49]
medical tinctures	$BrO_3^- + I^-$	[50]

phosphoric acid and naphthol is inhibited by molybdenum(VI). The decreased rate of this reaction is used for determination of molybdenum [31]. Several other reactions suitable for catalytic determination of molybdenum are listed in Table 11-3.

3. Applications to Specific Materials

Catalytic methods offer a convenient means for determining trace amounts of molybdenum without the need for expensive instrumentation. Precautions are, however, necessary to avoid interferences from other ions.

Table 11-4 lists some applications for determining molybdenum by catalytic procedures.

References

1. Yatsimirskii, K.B., Afanas'eva, L.P.: Zh. Anal. Khim., *11*, 319 (1956)
2. Ruzinov, L.P., Alekseeva, I.I., Khachaturyan, E.G.: Zh. Anal. Khim., *28*, 1109 (1973)
3. Erdey, L., Svehla, G.: Microchem. J. Symp. Ser., *2*, 343 (1962)
4. Sharipov, R.K., Songina, O.A.: Zavod. Lab., *29*, 1293 (1963)
5. Bulgakova, A.M., Mernaya, A.P.: Metody Anal. Khim. Reakt. Prep., (13), 143 (1966); Chem. Abstr., *69*, 92664v (1968)
6. Kambara, T., Tanaka, S., Fukada, K.: Bunseki Kagaku, *17*, 1144 (1968)
7. Pantel, S., Weisz, H.: Anal. Chim. Acta, *70*, 391 (1974)
8. Gaál, F.F., Sörös, V.I., Vajgand, V.J.: Anal. Chim. Acta, *84*, 127 (1971)
9. Altinata, A., Pekin, B.: Anal. Lett., *6*, 667 (1973)
10. Anokhina, L.G., Sopin, Yu.A., Agrinskaya, N.A.: Tr. Novocherk. Politekh. Inst., *143*, 63 (1963); Chem. Abstr., *61*, 27g (1964)
11. Svehla, G.: Analyst (London), *94*, 513 (1969)
12. Svehla, G., Erdey, L.: Microchem. J., *7*, 206 (1963)
13. Gaál, F.F., Sörös, V.I., Szebenyi, F.B., Canić, V.D.: Microchem. J., *22*, 535 (1977)
14. Weisz, H., Pantel, S.: Anal. Chim. Acta, *76*, 487 (1975)
15. Weisz, H., Rothmaier, K.: Anal. Chim. Acta, *82*, 155 (1976)
16. Hadjiioannou, T.P.: Anal. Chim. Acta, *35*, 360 (1966)
17. Weisz, H., Klockow, D., Ludwig, H.: Talanta, *16*, 921 (1969)
18. Weisz, H., Rothmaier, K.: Anal. Chim. Acta, *75*, 119 (1975)
19. Bradfield, E.G., Stickland, J.F.: Analyst (London), *100*, 1 (1975)
20. Quin, B.F., Woods, P.H.: Analyst (London), *104*, 552 (1979)
21. Alckseeva, I.I., Ruzinov, L.P., Khachaturyan, E.G., Chernyshova, L.M.: Zh. Anal. Chem., *35*, 60 (1980)
22. Feys, R., Dewynck, J., Tremillon, B.: Talanta, *22*, 17 (1975)
23. Kreingol'd, S.U., Vasnev, A.N.: Zavod. Lab., *44*, 265 (1978)
24. Sekheta, M.A., Milovanović, G.A., Janjić, T.J.: Mikrochim. Acta, I, 297 (1978)
25. Babko, A.K., Lisetskaya, G.S., Tsarenko, G.F.: Zh. Anal. Khim., *23*, 1342 (1968)
26. Pantaler, R.P.: Zh. Anal. Khim., *18*, 603 (1963)
27. Pavlova, V.K., Yatsimirskii, K.B.: Zh. Anal. Khim., *24*, 1347 (1969)
28. Pantel, S.: Anal. Chim. Acta, *104*, 205 (1979)
29. Jost, P., Lagrange, P., Wolff, C.M., Schwing, J.P.: J. Less Common Met., *36*, 169 (1974)

30. Wolff, C.M., Schwing, J.P.: Bull. Soc. Chim. Fr., 679 (1976)
31. Weisz, H., Vereno, I.: Anal. Chim. Acta, *91*, 229 (1977)
32. Yatsimirskii, K.B., Filippov, A.P.: Zh. Neorg. Khim., *9*, 2096 (1964)
33. Weisz, H., Meiners, W.: Anal. Chim. Acta, *90*, 71 (1977)
34. Klockow, D., Karenovics, G., Meiners, W.: Anal. Chim. Acta, *100*, 485 (1978)
35. Yatsimirskii, K.B., Filippov, A.P.: Zh. Anal. Khim., *20*, 815 (1965)
36. Kolosov, I.V., Kuz'mina, A.E.: Izv. Timiryazevsk. S'kh. Akad., (1), 203 (1969); Chem. Abstr., *70*, 91151j (1969)
37. Shiokawa, T.: J. Chem. Soc. Jpn. Pure Chem. Sect., *71*, 1 (1950)
38. Omarova, E.S., Speranskaya, E.F., Kozlovskii, M.T.: Izv. Akad. Nauk. Kaz. SSR, Ser. Khim., *18*(2), 32 (1968); Chem. Abstr., *69*, 32763s (1968)
39. Yamane, T., Kitamura, T., Fukasawa, T., Suzuki, T.: Bunseki Kagaku, *21*, 799 (1972)
40. Henrim, G., Scholz, F.: Z. Chem. (Leipzig), *20*, 150 (1980)
41. Thompson, H., Svehla, G.: Fresenius' Z. Anal. Chem., *247*, 244 (1969)
42. Dickson, E.L., Svehla, G.: Anal. Chim. Acta, *139*, 117 (1982)
43. Lazarev, A.I.: Zh. Anal. Khim., *22*, 1836 (1967)
44. Fuge, R.: Analyst (London), *95*, 171 (1970)
45. Christian, G.D., Patriarche, G.J.: Anal. Lett., *12B*, 11 (1971)
46. Jordanov, N., Pavlova, M., Stefanov, S.: Talanta, *25*, 389 (1978)
47. Cui, W., Yuan, X.: Fen Hsi Hua Hsueh, *9*, 461 (1981); Anal. Abstr., *42*, 3H80 (1982)
48. Wilson, A.M.: Anal. Chem., *38*, 1784 (1966)
49. Kuroda, R., Tarui, T.: Fresenius' Z. Anal. Chem., *269*, 22 (1974)
50. Pelczar, T., Janik, B.: Acta Pol. Pharm., *26*, 319 (1969); Chem. Abstr., *72*, 47423s (1970)

Chapter 12

Radiochemical and Activation Methods

There are seven naturally occurring isotopes of molybdenum and fifteen radioactive molybdenum isotopes. Only a few of the radioactive isotopes are suitable for analytical purposes. There application to determination of molybdenum as radioactive molybdenum, by isotopic dilution where a known amount of radioactive molybdenum is intentionally added to a sample, and in activation analysis is discussed in this chapter.

1. Radioactive Molybdenum Isotopes

1.1 Determining Radioactive Molybdenum

Radioactive ^{99}Mo, one of the products of uranium fission, is sometimes monitored to assess the quality of fissionable material and/or to indicate contamination of reactor coolant water [1,2]. It is also a source of ^{99m}Tc ($t_{1/2} = 6$ h). This isotope of technetium is used in nuclear medicine [3]. In addition to emitting its own radiation, e.g. γ-rays at 0.181 and 0.740 MeV [4], ^{99}Mo decays according to the following reactions [5]:

$$^{99}\text{Mo} + \text{n} \rightarrow\ ^{99m}\text{Tc} + \beta^- \qquad\qquad t_{1/2} = 66.48 \text{ h}$$
$$^{99m}\text{Tc} \rightarrow\ ^{99}\text{Tc} + \gamma \qquad\qquad\qquad t_{1/2} =\ 6.02 \text{ h}$$

Frequently technetium rather than molybdenum is monitored when determining the presence of molybdenum, thus helping to avoid interferences from other emitting nuclides within a fission sample. Procedures for separating ^{99}Mo from uranium fission products include sulfide precipitation [3], extraction with α-benzoinoxime [1,2], extraction with 2-hexylpyridine in the presence of thiocyanate ion [6], and ion exchange [7,8]. Frequently it is necessary to separate ^{99m}Tc from its parent ^{99}Mo. This is done using N-phenylbenzohydroxamic acid in chloroform to remove molybdenum while technetium remains in the aqueous phase [9] or by a ring-oven procedure which uses paper chromatography carried out on a heated, drilled metal block [10,11].

Following is the procedure of Maeck, Kussy, and Rein for *separation of* ^{99}Mo *from fission product mixtures* using extraction with α-benzoinoxime [2].

An aliquot of the fission mixture is placed in a 50 mL centrifuge tube. Add 3.00 mL of molybdenum carrier (containing exactly 18.5 g ammonium molybdate, $(NH_4)_6Mo_7O_{24} \cdot 4 H_2O$, dissolved in 100 mL of water plus 1 mL 0.5 M sodium bromate, $NaBrO_3$, and diluted to a final volume of exactly one liter with 6 M hydrochloric acid, HCl), 1 mL of 0.05 M potassium permanganate, $KMnO_4$, 30 mg iron(III) chloride, $FeCl_3$, 1 mL 1% (w/v) potassium iodide, KI, and distilled water to a final volume of 20 mL. Heat the solution to boiling and all dropwise 6 M ammonia, NH_3, solution until no further precipitate formation is observed. Centrifuge and decant the supernate into a 125 mL separatory funnel. Wash the precipitate with 5 mL of 1% (v/v) ammonia solution and add the washing to the separatory funnel. Add to the funnel 5 mL of 10 M HCl, 10 mL of 2% (w/v) α-benzoinoxime (in 95% ethanol), and 30 mL of ethyl acetate. Extract for two minutes and discard the aqueous layer. Wash the organic layer with 25 mL of 1 M HCl and discard the washing. Back extract the molybdenum by adding 15 mL 4 M NH_3. Extract for 5 min, until the organic layer is clear, and collect the aqueous layer. Add 10 mL of 10 M HCl to neutralize the ammonia in the aqueous layer and repeat the α-benzoinoxime extraction and back extraction.

The ammonical aqueous phase from the second back extraction is acidified with 5 mL of 15 M nitric acid, HNO_3. To precipitate molybdenum, add 3 mL of 5% (w/v) lead nitrate, $Pb(NO_3)_2$, solution. Add 6 M NH_3 dropwise until the pH is approximately 5. Add 1 mL of glacial acetic acid, $HC_2H_3O_2$. Centrifuge the precipitate. Discard the mother liqueur and collect the precipitate of lead molybdate, $PbMoO_4$, on a previously weighed one cm diameter filter disk (medium fine porosity) using water to transfer the precipitate from the centrifuge tube. Dry the $PbMoO_4$ for 10 min in an oven at 110 °C. Count the β or γ of ^{99}Mo. Weight the precipitate and correct the count for loss of carrier during the separation procedure.

1.2 Isotopic Dilution with Radioactive Molybdenum

In the isotopic dilution method of analysis, a known weight and activity of a radioactive isotope of molybdenum, generally ^{99}Mo or ^{101}Mo, is added to any nonradioactive molybdenum containing sample. Following chemical treatment to assure all the molybdenum present is in the same oxidation state, the total molybdenum content (that initially present plus that which has been added) is isolated by some appropriate means. This step need not be quantitative. The activity associated with the isolated molybdenum is then measured. By proportion one proceeds using the ratio:

$$\frac{\text{observed Mo activity}}{\text{added Mo activity}} = \frac{\text{wt. Mo isolated}}{\text{total wt. Mo in sample}}$$

The total weight of molybdenum present in the sample is composed of two terms, the weight of added radioactive molybdenum whose value is known and the original amount of molybdenum present whose value is sought. One solves for the value of sought molybdenum. Frequently the amount of added molybdenum, although perhaps exhibiting a very high activity, is of negligible weight. In this case the total weight of molybdenum calculated is the unknown weight of molybdenum.

A modification of the isotope dilution technique tags the reagent with which molybdenum is reacted rather than molybdenum itself. ^{110}Ag, for example, is used to precipitate molybdenum and the amount of silver in the precipitate monitored by its emission rather than by weighing Ag_2MoO_4 [12].

Using isotopic dilution Shamaev, Martynov, and Staroverova [13] with added ^{99}Mo but omitting additional molybdenum carrier, so the weight of added ^{99}Mo was negligible, complexed a molybdenum sample with EDTA in the presence of strong acid. Many interferring ions are not complexed by EDTA in strong acid although iron(III) and other metals strogly bound to EDTA interfere. In strong acid quantitative complexation between molybdenum and EDTA does not occur although the authors recommend at least a 30% complexation in the reaction of molybdenum and EDTA. The remaining, uncomplexed molybdenum is removed from the sample be precipitation with 8-hydroxyquinoline while leaving the Mo(VI)-EDTA complex intact. Following filtration, the filtrate is counted and a ratio established between counts of molybdenum observed, total counts added, amount of molybdenum in the filtrate (expressed in terms of the amount of added EDTA), and amount of molybdenum initially present. This latter value is the desired quantity. Originally developed for separation of molybdenum from tungsten, this procedure should be more widely applicable as the presence of various masking agents (tartrate, oxalate, and citrate) do not interfere with molybdenum determination [13]. Details of the procedure follow:

The unknown aliquot should contain about 10 mg Mo per 10 mL sample. Add an appropriate amount of carrier free $Na_2{}^{99}MoO_4$ to achieve the desired activity. Add dropwise either 1.0 M hydrochloric acid, HCl, or 1.0 M sodium hydroxide, NaOH, solution until the pH is approximately 5. Add 5 mL of 5% (w/v) hydrazine sulfate, $H_2NNH_2 \cdot H_2SO_4$ and sufficient distilled water until the final volume is about 50 mL. Boil the solution for 5 to 10 min. Cool the sample and add dropwise 1.0 M HCl until the pH is 2.9. Add 10 mL of buffer (containing 393 mL 0.1 M sodium citrate, $Na_3C_6H_5O_7$, and 607 mL of 0.1 M HCl). Add 5 mL of 0.1 M sodium tartrate, $Na_2C_4H_4O_6$, as a masking agent for tungsten. Add 4.0 mL of 0.010 M Na_2H_2EDTA solution. Heat the sample to near boiling and add dropwise with stirring a 10% (w/v) solution of 8-hydroxyquinoline (in ethanol) until no further precipitate formation is observed. Approximately 1 to 1.5 mL should be sufficient to precipitate the uncomplexed molybdenum. Allow the sample to remain near boiling for a few minutes and then filter through a fine (number 3) sintered glass crucible. Collect the filtrate in a 100 mL volumetric flask. Wash the precipitate once or twice with 5 mL portions of the sodium citrate-hydrochloric acid buffer, collecting the washings along with the filtrate. Dilute the filtrate and washings to exactly 100 mL with distilled water. Remove and count a suitable portion of the solution to determine the ^{99}Mo activity.

$$\frac{\text{observed Mo activity}}{\text{added Mo activity}} = \frac{2 \times \text{mmol EDTA added}}{\text{mmole Mo in sample}}$$

The factor 2 is necessary because the Mo:EDTA ratio is 2:1. Solve the expression for mmole Mo in the sample. Convert this value to mg Mo and correct, if necessary, for whatever dilutions were made in choosing an aliquot of the original sample.

2. Activation Analysis

Activation analysis, in which artificially produced radioisotopes are formed within a sample by bombardment from externally generated atomic particles, is a well established technique for simultaneous multicomponent analysis. Irradiation, generally with neutrons, protons, or alpha particles, results in radioactive nuclides of all components present. By selective monitoring of the decay of these isotopes, a complete analysis of a sample is possible. Irradiation can occur from several minutes to several days depending upon the source, the sample, and the elements under study. Counting, generally of induced gamma radiation [14], begins anywhere from several minutes to several days following irradiation depending upon the half-lives of the components formed. If, for example, interferring elements have short half-lives, it is better to wait a few days until they have decayed before conting. This removes their possible interference. With thermal neutron activation ^{99}Mo ($t_{1/2} = 66.5$ h) and ^{101}Mo ($t_{1/2} = 14.6$ min) are formed from the most abundant naturally occurring molybdenum isotopes ^{98}Mo (natural abundance 23.8%) and ^{100}Mo (natural abundance 9.6%) respectively. The daughter products ^{99m}Tc ($t_{1/2} = 6$ h) and ^{101}Tc ($t_{1/2} = 14$ min) are sometimes monitored as are other possible molybdenum isotopes [15]. In addition to thermal neutron bombardment, epithermal neutrons, protons, and alpha particles are all used as sources for activation analysis. In each case interaction with different molybdenum isotopes is favored and different products result. The 0.74 MeV γ-radiation of ^{99}Mo may not, for example, be the optimum choice for following molybdenum decay with these sources [16–20].

Matrix effects are most important in any activation analysis procedure. Standards containing approximately equivalent but known amounts of molybdenum and a matrix of approximately the same composition should be analyzed along with the unknown molybdenum containing sample. One must be careful that interferring emissions from other activated elements are not mistaken for molybdenum. The ^{99m}Tc emission at 0.140 MeV is very close to the 0.143 MeV ^{52}Fe and 0.134 ^{187}W radiations, for example [21]. Schemes in which the separation of molybdenum from various interferences following activation is specifically mentioned include precipitation with Pb(II) [22], precipitation with nitron [23] and extraction with diethyldithiocarbamate in chloroform [24]. Separation of molybdenum from specific matrices as part of an activation procedure include separation in ores by adsorption chromatography [25], ion exchange [26], fusion [27], and extraction [28]; in steels by extraction [29]; in tungsten [30] and uranium [31] by ion exchange; and in biological samples by adsorption chromatography [32], ion exchange [33], and extraction [34].

3. Application to Specific Materials

Radiochemical and activation methods are powerful analytical techniques. Often they provide answers to questions not easily answered by other means. The ability to monitor radioactive tracers in living systems, for example, is now common in medical diagnostics. The nondestructive nature of activation analysis is applied to rare and irreplaceable art objects. This, coupled with the extreme sensitivity of activation techniques, frequently at the ng/mL level, make this form of analysis extremely useful provided, of course, a suitable activation source is available to irradiate the sample. Sources for activation analysis are expensive and require skilled operators in addition to consideration of health and safety aspects. All radiochemical techniques require special precautions to assure the health and safety of the experimenter. Table 12-1 lists representative procedures for molybdenum in various samples using radiochemical and activation procedures.

Table 12-1. Radiochemical and Activation Analysis of Molybdenum in Specific Materials

Material	Procedure	Ref.
rocks and minerals		
silicates	neutron activation, ^{99m}Tc, $t_{1/2}$ 6 h, γ at 0.14 MeV	[26]
sandstone	alpha activation, ^{97}Ru, $t_{1/2}$ 2.9 day, γ at 0.22 MeV	[27]
cassiterite	neutron activation, ^{99m}Tc, $t_{1/2}$ 6 h, γ at 0.14 MeV	[25]
Cu-Mo ore	neutron activation, ^{99m}Tc, $t_{1/2}$ 6 h, γ at 0.14 MeV	[35]
scheelite	neutron activation, ^{99m}Tc, $t_{1/2}$ 6 h, γ at 0.14 MeV	[28]
ferrous alloys		
cast iron	neutron activation, ^{99m}Tc, $t_{1/2}$ 6 h, γ at 0.14 MeV	[29]
low alloy steel	neutron activation, ^{99m}Tc, $t_{1/2}$ 6 h, γ at 0.14 MeV	[36]
stainless steel	neutron activation, ^{91}Mo, $t_{1/2}$ 15.5 min, γ at 0.51 MeV	[37]
Cr-Mo-W-Co steel	alpha activation, ^{95m}Tc, $t_{1/2}$ 61 day, γ at 0.20 MeV	[38]
W-high speed steel	alpha activation, ^{95m}Tc, $t_{1/2}$ 61 day, γ at 0.20 MeV	[39]
steel	isotopic dilution, add ^{99}MoO$_3$, count β and γ	[40]
steel	neutron activation, ^{99}Mo, $t_{1/2}$ 66 h, γ at 0.74 MeV	[41]
pure elements		
aluminium	neutron activation, ^{99m}Tc, $t_{1/2}$ 6 h, γ at 0.14 MeV	[42, 43]
beryllium	neutron activation, ^{99}Mo, $t_{1/2}$ 66 h, γ at 0.74 MeV	[44]
cobalt	proton activation, ^{94m}Tc, $t_{1/2}$ 53 min, γ at 0.87 MeV	[45]
chromium	neutron activation, ^{99}Mo, $t_{1/2}$ 66 h, γ at 0.74 MeV	[46]
iron	neutron activation, ^{99m}Tc, $t_{1/2}$ 6 h, γ at 0.14 MeV	[47]
niobium	neutron activation, ^{101}Mo, $t_{1/2}$ 14.6 min, γ at 0.19 MeV	[48]
selenium	neutron activation, ^{99m}Tc, $t_{1/2}$ 6 h, γ at 0.14 MeV	[49]
tantalum	proton activation, ^{94}Tc, $t_{1/2}$ 4.8 h, γ at 0.70 MeV	[50, 51]
tellurium	neutron activation, ^{99m}Tc, $t_{1/2}$ 6 h, γ at 0.14 MeV	[49]
tungsten	isotopic dilution, add ^{99}Mo, precipitate ^{99}MoS$_3$	[52]

Table 12-1 (continued)

Material	Procedure	Ref.
uranium	neutron activation, ^{99m}Tc, $t_{1/2}$ 6 h, γ at 0.14 MeV	[31]
zirconium	neutron activation, ^{99}Mo, $t_{1/2}$ 66 h, γ at 0.74 MeV	[53]
biological samples		
animal blood	neutron activation, ^{99m}Tc, $t_{1/2}$ 6 h, γ at 0.14 MeV	[54, 55]
human serum	neutron activation, ^{99m}Tc, $t_{1/2}$ 6 h, γ at 0.14 MeV	[56]
hair	neutron activation, ^{101}Tc, 13 min, γ at 0.31 MeV	[57]
liver	neutron activation, ^{99}Mo, $t_{1/2}$ 66 h, γ at 0.74 MeV	[58, 59]
bone	isotopic dilution, add Na$_2$ ^{99}MoO$_4$	[34]
teeth	neutron activation, ^{101}Mo, $t_{1/2}$ 14.6 min, γ at 0.19 MeV	[60]
orchard leaves	neutron activation, ^{101}Mo, $t_{1/2}$ 14.6 min, γ at 0.19 MeV	[61]
cotton	neutron activation, ^{101}Mo, $t_{1/2}$ 14.6 min, γ at 0.19 MeV	[62]
tobacco	neutron activation, ^{99m}Tc, $t_{1/2}$ 6 h, γ at 0.14 MeV	[63]
wheat flour	neutron activation, ^{99m}Tc, $t_{1/2}$ 6 h, γ at 0.14 MeV	[64]
plants	neutron activation, ^{99}Mo, $t_{1/2}$ 66 h, γ at 0.74 MeV	[65]
soil	neutron activation, ^{99m}Tc, $t_{1/2}$ 6 h, γ at 0.14 MeV	[66]
soil	neutron activation, ^{99}Mo, $t_{1/2}$ 66 h, γ at 0.74 MeV	[67]
coal	neutron activation, ^{99}Mo, $t_{1/2}$ 66 h, γ at 0.74 MeV	[44]
environmental samples		
sea water	neutron activation, ^{99}Mo, $t_{1/2}$ 66 h, γ at 0.74 MeV	[68, 69]
catalysts		
MoS$_2$ containing catalyst	isotopic dilution, add ^{35}S, precipitate BaSO$_4$	[70]
hydrodesulfurization catalyst	neutron activation, ^{99m}Tc, $t_{1/2}$ 6 h, γ at 0.14 MeV	[71]

References

1. Scadden, E.M.: Nucleonics, *15*, 102 (1957)
2. Maeck, W.J., Kussy, M.E., Rein, J.E.: Anal. Chem., *33*, 237 (1961)
3. Tanase, M., Kase, T., Shikata, E.: J. Nucl. Sci. Technol., *13*, 591 (1976)
4. Bereznai, T.: Fresenius' Z. Anal. Chem., *302*, 353 (1980)
5. Alvarez, J., Cortés, F., Pérez, T.: J. Radioanal. Chem., *52*, 471 (1979)
6. Iqbal, M., Ejaz, M.: J. Radioanal. Chem., *47*, 25 (1979)
7. Dupuis, M.C., Dupuis, M.: Radiochim. Acta, *2*, 4 (1963)
8. Markl, I., Bobleter, O.: Fresenius' Z. Anal. Chem., *219*, 160 (1966)
9. Mikulaj, V., Macasek, F., Steinerova, M.: Radiochem. Radioanal. Lett., *29*, 199 (1977)
10. Hilton, D.A., Reed, D.: Analyst (London), *89*, 599 (1964)
11. Klockow, D.: Talanta, *14*, 817 (1967)
12. Govaerts, J., Barcia-Goyanes, C.: Nature, *168*, 198 (1951)
13. Shamaev, V.I., Martynov, A.M., Staroverova, L.L.: Zh. Anal. Khim., *27*, 2058 (1972)
14. Kusaka, Y., Tsuji, H., Fujii, I., Muto, H., Miyoshi, K.: Bull. Chem. Soc. Jpn., *38*, 616 (1965)
15. Osterhage, W.W.: J. Radioanal. Chem., *56*, 267 (1980)
16. Perdijon, J.: Anal. Chem., *39*, 448 (1967)
17. Krivan, V.: Anal. Chim. Acta, *79*, 161 (1975)
18. Kormali, S.M., Swindle, D.L., Schweikert, E.A.: J. Radioanal. Chem., *31*, 437 (1976)
19. Bäuerle, W., Krivan, V., Münzel, H.: Anal. Chem., *48*, 1434 (1976)
20. Sastri, C.A., Petri, H., Erdtmann, G.: Anal. Chem., *49*, 1510 (1977)
21. Steinnes, E.: Anal. Chem., *48*, 1440 (1976)
22. Thompson, B.A.: Anal. Chem., *31*, 1492 (1959)
23. Quaim, S.M., Butement, F.D.S.: Anal. Chim. Acta, *28*, 591 (1963)
24. Wyttenback, A., Bajo, S.: Anal. Chim. Acta, *47*, 2 (1975)
25. Brandone, A., Meloni, S., Girardi, F., Sabbioni, E.: Analusis, *2*, 300 (1973)
26. Gladney, E.S.: Anal. Lett., *11A*, 429 (1978)
27. Artem'ev, O.I., Kiselev, B.G., Pozdnyakov, S.V.: Zh. Anal. Khim., *34*(11), 2227 (1979)
28. Alian, A., Born, H.J., Stark, H.: Radiochim. Acta, *18*, 50 (1972)
29. Thompson, B.A. La Fleur, P.D.: Anal. Chem., *41*, 852 (1969)
30. Grosse-Ruyken, H., Döge, H.G.: Talanta, *12*, 73 (1965)
31. Kosta, L., Cook, G.B.: Talanta, *12*, 977 (1965)
32. Malvano, R., Grosso, P., Zanardi, M.: Anal. Chim. Acta, *41*, 251 (1968)
33. Nadkarni, R.A., Morrison, G.H.: Anal. Chem., *50*, 294 (1978)
34. Healy, W.B., McCabe W.J.: Anal. Chem., *35*, 2117 (1963)
35. Nikitin, V.N., Mordasov, V.R., Steblich, L.E., Chepurnoi, Yu.A., Shchetinin, A.M.: Zavod. Lab., *43*, 547 (1977)
36. Thompson, B.A., La Fleur, P.D.: Anal., *41*, 1888 (1969)
37. Gangadharan, S., Yegnasubramanian, S., Misra, S.C., Gupta, U.C.: J. Radioanal. Chem., *24*, 57 (1975)
38. Gihwala, D., Giles, I.S., Peisach, M.: J. Radioanl. Chem., *47*, 145 (1978)
39. Gihwala, D., Peisach, M.: J. Radioanal. Chem., *55*, 163 (1980)
40. Geldhof, M.L., Eeckhaut, J., Cornand, P.: Bull. Soc. Chim. Belg., *65*, 706 (1956)
41. Nadharni, R.A., Haldar, B.C.: Talanta, *16*, 116 (1969)
42. Yakovlev, Yu.V., Stepanets, O.V., Savel'ev, B.V.: Zh. Anal. Khim., *31*, 1215 (1976)
43. Lo, J.G., Ke, C.N., Tanaka, S., Yeh, S.J.: J. Chim. Chem. Soc. (Taipei), *24*, 21 (1977); Chem. Abstr., *87*, 94968t (1977)
44. Tamura, N.: Radiochem. Radioanal. Lett., *18*, 135 (1974)
45. Krivan, V.: Talanta, *23*, 621 (1976)

46. Loos-Neskovic, C., Fedoroff, M., Revel, G.: Anal. Chim. Acta, *85*, 95 (1976)
47. De Wispelaere, C., Op de Beeck, J.P., Hoste, J.: Anal. Chim. Acta, *70*, 1 (1974)
48. Faix, W.G., Caletka, R., Krivan, V.: Anal. Chem., *53*, 1594 (1981)
49. Shamaev, V.I.: Radiokhimiya, *2*, 624 (1960); Chem. Abstr., *56*, 11b (1962)
50. Barrandon, J.N., Benaben, P., Debrun, J.L., Valladon, M.: Anal. Chim. Acta, *73*, 39 (1974)
51. Krivan, V., Swindle, D.L., Schweikert, E.A.: Anal. Chem., *46*, 1626 (1974)
52. Shamaev, V.I., Martynov, A.M., Staroverova, L.L.: Zh. Anal. Khim., *27*, 2260 (1972)
53. Miklishanskii, A.Z., Leikin, Yu.A., Yakovlev, Yu.V., Savel'ev, B.V.: Zh. Anal. Khim., *29*, 1284 (1974)
54. Weers, C.A., Van der Sloot, H.A., Das, H.A.: J. Radioanal. Chem., *20*, 529 (1974)
55. Bagdavadze, N.V., Mosulishvili, L.M.: J. Radioanal. Chem., *24*, 65 (1975)
56. Maziere, B., Gros, J., Comar, D.: J. Radioanal. Chem., *24*, 279 (1975)
57. Healy, W.B., Bate, L.C.: Anal. Chim. Acta, *33*, 443 (1965)
58. Brune, D., Bivered, B.: Anal. Chim. Acta, *85*, 411 (1976)
59. Tjioe, P.S., DeGoeij, J.J.M., Houtman, J.P.W.: J. Radioanal. Chem., *37*, 511 (1977)
60. Livingston, H.D., Smith, H.: Anal. Chem., *39*, 538 (1967)
61. Diksic, M., Cole, T.F.: Anal. Chim. Acta, *93*, 261 (1977)
62. Rustamov, R., Khatamov, Sh., Orestova, I.I., Kist, A.A.: At. Energ., *34*, 476 (1973); Chem. Abstr., *82*, 15614v (1975)
63. Wyttenback, A., Bajo, S., Häkkinen, A.: Beitr. Tabakforsch., *8*, 247 (1976); Chem. Abstr., *85*, 119621y (1976)
64. Zmijewska, W.: J. Radioanal. Chem., *58*, 367 (1980)
65. Neuberger, M., Fourcy, A.: J. Radioanal. Chem., *1*, 289 (1968)
66. Zmijewska, W., Minczewski, J.: Chem. Anal. (Warsaw), *14*, 23 (1969)
67. Korshunov, Yu.F., Zhuk, L.I., Orestova, I.I., Gureev, E.S., Kist, A.A.: Zh. Anal. Khim., *31*, 1962 (1976)
68. Fujinaga, T., Kusaka, R., Koyama, M., Tsuji, H., Mitsuji, T., Imai, S., Okuda, J., Takamatsu, T., Ozaki, T.: J. Radioanal. Chem., *13*, 301 (1973)
69. Kulathilake, A.I., Chatt, A.: Anal. Chem., *52*, 828 (1980)
70. Todorovskii, D., Todorov, K., Kostadinov, K.: Isotopenpraxis, *15*, 290 (1979); Anal. Abstr., *39*, 6B115 (1980)
71. Heurtebise, M., Buenafama, H., Lubkowitz, J.A.: Anal. Chim., *48*, 1969 (1976)

Chapter 13

Rocks and Mineral Samples

Trace amounts of molybdenum are widely distributed in various rocks, minerals, and elsewhere throughout the soils and waters of the earth. Molybdenite, MoS_2, is the principle molybdenum ore but even here the concentration of molybdenum is not high. A second source of molybdenum is from the mining of copper and tungsten ores with which it is frequently associated. Chief production of molybdenum is in the United States, Canada, the Union of Soviet Socialist Republics, and the People's Republic of China with lesser amounts produced elsewhere [1]. Common molybdenum containing minerals are listed in Table 13-1.

Molybdenum is included in a general discussion of trace metal analysis of silicates [4] and other geological materials [5, 6]. *Rock samples* are placed in solution using either acid digestion or fusion.

Schweizer [7]: Five mL 40% (w/v) hydrofluoric acid, HF, plus three mL 70% (w/v) perchloric acid, $HClO_4$, is added to 0.5 g powdered sample in a teflon beaker. After dissolving, evaporate the sample to dryness using a sand bath. Repeat the acid treatment with 5 mL $HF + 3$ mL $HClO_4$ and again evaporate to dryness. Add 5 mL 12 M hydrochloric acid, HCl. Heat gently and evaporate the HCl. Dissolve the residue in 1.2 M HCl. Filter the solution through a fine filter paper (e.g. Whatman number 42). Wash the paper with additional 1.2 M HCl, collect the filtrate and washings, and dilute the sample to a known volume with additional 1.2 M HCl.

Table 13-1. Principle Molybdenum Minerals [2, 3]

Mineral	Composition
belonesite	$MgMoO_4$
chillagite	$3PbWO_4 \cdot PbMoO_4$
eosite	$3PbO \cdot V_2O_4 \cdot MoO_3$
ferrimolybdite	$Fe_2(MoO_4)_3 \cdot 8H_2O$
ilsemannite	$MoO_2 \cdot 4MoO_3$
jordisite	$MoO_3 \cdot SO_3 \cdot 5H_2O$
koechlinite	Bi_2MoO_6
lindgrenite	$Cu_3(MoO_4)_2(OH)_2$
molybdenite	MoS_2
pateraite	$CoMoO_4$
powellite	$CaMoO_4$
wulfenite	$PbMoO_4$

Kawabuchi and Kuroda [8]: Ten mL 18 M sulfuric acid, H_2SO_4, plus four mL 15 M nitric acid, HNO_3, plus ten mL 40% (w/v) hydrofluoric acid, HF, is added to 2 g powdered sample in a platinum dish. Heat the mixture on a hot plate and after the sample has dissolved, evaporate the liquid to fumes of sulfur trioxide, SO_3. Add 2 mL 18 M H_2SO_4 and 20 mL water to the residue and again fume the sample until SO_3 appears. Again add 2 mL 18 M H_2SO_4 and 20 mL water. Heat the mixture to near boiling for 30 min. Dilute the sample with water and proceed with the determination.

Sandell [9]: Mix 1 g finely powdered sample with 0.5 g anhydrous sodium carbonate, Na_2CO_3, in a platinum crucible. Subject the covered crucible to the full heat of a burner for 20–30 min. After cooling, add 2–3 mL water and heat the crucible gently to loosen the residue. Transfer the residue to a 150 mL beaker, add 2–3 drops ethanol, C_2H_5OH, and 30–40 mL water. Heat the mixture to near boiling until the residue dissolves. Filter the solution through a fine filter paper (e.g., Whatman number 42). Wash the residue 4–5 times with hot 1% (w/v) Na_2CO_3. Collect the filtrate and washings for further treatment.

Ward [10]: 0.1 g of finely ground sample is mixed with 0.5 g fusion material (anhydrous sodium carbonate, Na_2CO_3 – potassium nitrate, KNO_3, 1:1 wt/wt) in a heavy wall test tube. The sample is heated over an open flame for about 5 min. Add 4 mL water to the cooled melt and place the tube in a boiling water bath for 3–5 min. Filter the solution through a fine filter paper (e.g. Whatman number 42). Filtrate and washings are collected for further treatment.

Molybdenum ores and concentrates are also dissolved by acid digestion [11–13] and fusion [14, 15].

Sharipova and Songina [16]: 0.5 g finely ground sample is placed in a 150 mL beaker and 20 mL 15 M nitric acid, HNO_3, added. Heat the mixture until dissolved. Add 10 mL 9 M sulfuric acid, H_2SO_4, and evaporate the sample until fumes of sulfur trioxide, SO_3, appear. After cooling, dissolve the residue in water, heating until solution is complete.

Hope [17]: Place 15 g sodium hydroxide, NaOH, in an iron crucible and heat until the material melts. Allow the NaOH to cool and store the crucible in a desiccator until ready for use. Add 1 g finely ground sample to the anhydrous NaOH and again with a flame fuse the mixture. Cool the crucible by partially immersing it in a container of cold water. Transfer the crucible to a 250 mL beaker containing 80 mL water. Cover the beaker and heat to dissolve the melt. Rinse the crucible with hot 5% (w/v) NaOH and filter the sample using a course filter paper (e.g., Whatman number 41). Wash the residue with additional hot 5% (w/v) NaOH collecting the filtrate and washings. Acidify the sample and preceed with molybdenum determination.

Budesinsky [18]: A fusion mixture is prepared from 900 g sodium peroxide, Na_2O_2, and 300 g sodium hydroxide, NaOH. Place 8.0 g fusion mixture and 0.5 g sample in an iron or zirconium crucible. Mix the components well and heat in a furnace at 660 °C for 8–10 min. Cool the crucible by partially immersing it in cold water. When cool, place the crucible in a 200 mL beaker and wash the contents from the crucible with three 15 mL portions of distilled water. Rinse and remove the crucible. Add 1 mL ethanol, C_2H_5OH, and heat the sample to near boiling. After cooling, acidify the sample and continue with the determination of molybdenum.

Pollock [19]: Mix 1 g finely powdered sample and 8 g sodium peroxide, Na_2O_2, in a zirconium crucible. Cover the mixture with 2 g additional Na_2O_2. Place a lid on the crucible and slowly apply heat from an open flame until the crucible is dull red. Cool the crucible and place it in a 250 mL beaker. Add 40 mL water and 10 mL 12 M Hydrochloric acid, HCl. Heat to dissolve the fused sample. Remove and rinse the crucible and lid before proceeding with the molybdenum determination.

Care is needed in determining molybdenum in geological samples. Atomic absorption spectrometry, neutron activation analysis, and x-ray fluorescence analysis

Table 13-2. Determination of Diverse Elements in Molybdenum Ores and Concentrates

Element	Procedure	Ref.
Al, Bi, Ca, Fe, Mg, Pb, Si	emission spectroscopy	[22]
Au, Pd	atomic absorption	[23]
Cu, Fe, Pb	compleometric titration	[24]
Cu, Zn	polarography	[25]
Ge	colorimetry phenylfluorone	[26]
Na	emission spectroscopy	[27]
Nb	colorimetry thiocyanate	[28]
Pb	precipitation chromate	[29]
Re	colorimetry furil α-dioxime	[18, 30]
Re	colorimetry 1-phenyl-2-thiourea	[19]
Re	colorimetry safranine T	[31]
Re	colorimetry thiocyanate	[32, 33]
Re	emission spectroscopy	[34]
Re	polarography	[14, 35]
Re	neutron activation analysis	[36]
Sn	colorimetry phenylfluoronc	[37]
W	neutron activation analysis	[38]
W	colorimetry thiocyanate	[39]

for molybdenum in waste water from uranium tailings, for example, produced divergent results with spiked samples [20]. Errors in analysis of molybdenum concentrates have recently been discussed [21].

Numerous diverse ions are found in molybdenum containing ores and minerals. Methods for determining some of these are listed in Table 13-2.

Donaldson has successfully determined molybdenum in various standard ore samples by atomic absorption spectrometry [13]. Molybdenum concentration ranged from $<0.01\%$ to $\sim 0.6\%$ in the ores studied. After dissolving the sample in nitric acid, molybdenum is extracted with α-benzoinoxime. If excessive amounts of tungsten are present, a second extraction, using ethyl xanthate, is necessary before absorbance measurements are taken. Samples analyzed included Zn-Sn-Cu-Pb ore, Cu-Mo ore, Mo ore, and W ore. Following are details of the procedure.

A calibration curve is prepared from known molybdenum solutions by dissolving 1.500 g molybdenum trioxide, MoO_3, in 50 mL 2% (w/v) sodium hydroxide, NaOH. Transfer the solution to a one liter volumetric flask and add sufficient water to obtain exactly one liter. This solution contains 1,000 µg Mo/mL. Transfer exactly 25.0 mL of the solution to a 250 mL volumetric flask and dilute to 250 mL with water. The dilute stock solution contains 100 µg Mo/mL. Transfer known amounts from 1.00 mL dilute stock solution (100 µg Mo) to 30.00 mL (3,000 µg Mo) to each of several 100 mL beakers. To each beaker add 10 drops 50% (v/v) sulfuric acid, H_2SO_4. Carefully evaporate each sample to dryness. Cool, rinse the sides of each beaker with water, and again evaporate each sample to dryness. After cooling add to each sample 1 mL 25% (v/v) ammonia, NH_3, one drop 0.2% (w/v) phenolphthalein (ethanol), and 5 mL water. Heat each sample until colorless, pH <9. Add to each sample 10 mL 12 M hydrochloric acid, HCl, and 10 mL 1.0% (w/v) aluminum solution. (Prepare by dissolving 10 g aluminum metal in 400 mL 50% (v/v) HCl. Filter if necessary (fine filter paper, e.g. Whatman number 42) and add 300 mL 12 M HCl. Now add sufficient water to make a final volume of one liter.) Quantitatively transfer the contents of each beaker to a separate 100 mL volumetric flask. Add water to each solution until the final volume of each is exactly 100 mL. An additional flask containing only 10 mL 12 M HCl and 10 mL 1.0% (w/v) aluminum solution serves as the blank, zero molybdenum standard. The standard prepared from 1.00 mL diluted stock solution contains 1.00 μg Mo/mL and that prepared from 30.00 mL diluted stock solution contains 30.00 μg Mo/mL. The treatment of standards described is necessary to assure that molybdenum in the standards is subjected to the same treatment as molybdenum in the unknown sample. Molybdenum absorbance is measured at 313.3 nm with a lamp current of 5 mA. Spectral band pass is 0.1 nm and the burner height is 8 mm. A fuel-rich, brightly luminous flame is used with acetylene flow approximately 3 L/min and air flow approximately 13 L/min. Exact instrument settings near these values are achieved by aspirating the most concentrated molybdenum standard and fine adjustment of the controls to obtain maximum absorbance. Once this is done, the blank and each standard are measured in turn and a calibration curve prepared of absorbance reading versus molybdenum concentration.

A finely powdered ore sample, between 0.1 and 1.0 g and containing up to 2.5 mg Mo, is placed in a 400 mL teflon beaker. No more than 25 mg tungsten should be present in the sample. Add 20 mL 50% (v/v) nitric acid, HNO_3, and 5 mL 20% (v/v) bromine, Br_2, (CCl_4). Cover the beaker and allow the sample to stand for 10 min. Heat the sample on a sand bath to remove Br_2 and CCl_4. Add 10 mL 12 M HCl and 20 mL 50% (v/v) H_2SO_4. Heat the sample until evolution of nitrogen oxides ceases. Remove the cover, rinse the sides of the beaker with water, and add 5 mL 40% (v/v) hydrofluoric acid, HF. Heat carefully until the sample volume is reduced to approximately 3 mL. Cool the sample, add 25 mL water, and heat to dissolve the residue. Add 10 mL 20% (v/v) Br_2 (CCl_4) solution and heat until excess bromine and CCl_4 are removed. Add 10 mL 20% (w/v) tartaric acic solution, $C_4H_6O_6$. Add in 0.5 mL increments 50% (w/v) sodium hydroxide, NaOH, until the sample is basic to litmus. After standing several minutes add 12 M HCl dropwise until the solution is acidic. Add an additional 15 mL 12 M HCl and if a precipitate persists, filter the solution through a course filter paper (e.g. Whatman number 41). Filtrate and washings are collected in a 250 mL separatory funnel. Discard any residue remaining on the filter paper. Add to the sample in the separatory funnel 5 mL of freshly prepared 10% (w/v) iron(II) ammonium sulfate, $FeSO_4 \cdot (NH_4)_2SO_4$. Add water to the funnel until the final sample volume is 100 mL.

Add 15 mL 0.2% (w/v) α-benzoinoxime, $C_{14}H_{13}O_2N$, ($CHCl_3$). Shake the funnel for 1 min and allow several minutes for the layers to separate. Drain the $CHCl_3$ layer and any solid complex that forms into a 150 mL beaker. Repeat the extraction three more times, a total of 60 mL α-benzoinoxime solution. Combine all extracts. If the sample contains less than 2 mg tungsten, one proceeds directly with the molybdenum determination. (Should the molybdenum concentration itself be low, less than 2 mg tungsten can interfere.) Add 10 mL 50% (v/v) HNO_3 to the combined chloroform extracts and heat on a water bath to remove $CHCl_3$. Add 3 mL 70% (w/v) perchloric acid, $HClO_4$, and 2 mL 50% (v/v) H_2SO_4. Cover the beaker and heat to expel nitrogen oxides. Remove the cover and carefully evaporate the sample to dryness. Wash the sides of the beaker with water and again evaporate the sample to dryness.

If excessive amounts of tungsten are present in the sample, yellow tungsten(VI) oxide, WO_3, insoluble in water, is observed at this point. Should this be the case, add 0.5 mL 70% (w/v) $HClO_4$ and evaporate the sample to dryness. Add 25 mL water, 5 mL 20% (w/v) tartaric acid, and 1 drop 0.2% (w/v) phenolphthalein. Add slowly 50% (w/v) NaOH until the solution turns red, pH > 9. Follow this with dropwise addition of 12 M HCl until the solution is colorless, then add 6 mL more. Quantitatively transfer the solution to a 125 mL separately funnel, make up to 50 mL volume with water, and proceed with the ethyl xanthate extraction.

If no excess of tungsten is observed, the sample residue is treated in a manner analogous to the standard samples by heating with 1 mL 25% (v/v) NH_3, 1 drop 0.2% (w/v) phenolphthalein, and 3 mL water. Continue heating until the solution is colorless, pH < 9. If the amount of molybdenum is low use a 10 mL volumetric flask otherwise use a 100 mL volumetric flask. Add either 1 mL or 10 mL 12 M HCl depending upon the final sample volume. Follow this with either 1 mL or 10 mL 1.0% (w/v) aluminum solution, again depending upon the final sample volume. Transfer the sample to the volumetric flask and dilute to the final volume with water. Mix the contents of the flask thoroughly. Aspirate the sample into the flame and record the absorbance. From the calibrations graph, determine the amount of molybdenum present, correcting if necessary for the size of the volumetric flask.

If tungsten interference is known to be present, the chloroform extracts of molybdenum with α-benzoinoxime are collected and placed in a clean 125 mL separatory funnel. Add 10 mL 15 M NH_3 and back extract molybdenum into the ammonia solution, shaking the funnel for 4 min. Wait 5 min for the layers to separate and discard the chloroform layer. Add to the basic solution 5 mL 20% (w/v) tartaric acid, 20 mL water, and 15 mL 12 M HCl. Blow ammonium chloride, NH_4Cl, fumes from the funnel with a stream of air and cool the sample to room temperature. Add 2 mL freshly prepared 20% (w/v) potassium ethyl xanthate, $C_3H_5OS_2K$. Use caution as exposure to xanthate vapor can be hazardous. Continue the extractions in a fume hood. Mix the contents of the funnel thoroughly and allow the mixture to stand for about one minute to form the colored molybdenum-ethyl xanthate complex. Add 10 mL $CHCl_3$ and shake the funnel for one minute. Wait several minutes for the layers to separate before transferring the $CHCl_3$ layer to a 150 mL beaker. Extract the aqueous layer three more times using 1.0, 0.5, and 0.5 mL ethyl xanthate with 10, 5, and 5 mL $CHCl_3$ respectively. If after the fourth extraction the aqueous phase is still pink, continue extracting using 0.5 mL ethyl xanthate with 5 mL $CHCl_3$ until both phases are colorless. Add 10 mL 50% (v/v) HNO_3 to the combined xanthate extracts and heat in a hot water bath to remove $CHCl_3$. Add 10 mL 70% (w/v) $HClO_4$ and 5 mL 50% (v/v) H_2SO_4. Cover the beaker and heat to expel nitrogen oxides. Remove the cover and carefully evaporate the sample to dryness. Wash the sides of the beaker with water and again evaporate the sample to dryness. Continue with the addition of 25% (v/v) NH_3, pH adjustment, and dilution to a known volume as described earlier following the α-benzoinoxime extraction in the absence of tungsten. The sample is aspirated into the flame and the amount of molybdenum present read from the calibration graph at the measured absorbance value.

References

1. Kummer, J.T.: U.S. Bur. Mines, Bull., No. 671 (1980)
2. Browning, P.E.: Introduction to the Rarer Elements, 1917, New York, John Wiley
3. Killeffer, D.H., Lenz, A.: Molybdenum Compounds, 1952, New York, Wiley-Interscience
4. Strelow, F.W.E., Liebenberg, C.J., Toerien, F. von S.: Acta Chim. Acta, *47*, 251 (1969)
5. Blyum, I.A., Zolotov, Yu.A.: Zh. Anal. Khim., *31*(1), 159 (1976)
6. Steger, H.F.: Talanta, *23*, 81 (1976)
7. Schweizer, V.B.: At. Absorpt. Newsl., *14*, 137 (1975)
8. Kawabuchi, K., Kuroda, R.: Talanta, *17*, 67 (1970)
9. Sandell, E.B.: Ind. Eng. Chem., Anal. Ed., *8*, 336 (1936)
10. Ward, F.N.: Anal. Chem., *23*, 788 (1951)
11. Yurkevich, Yu.N., Shapiro, K.Ya.: Met. Vol'frama Molibdena Niobiya, *53* (1967); Chem. Abstr., *69*, 98546e (1968)
12. Fed'kovskii, I.A.: Zavod, Zab., *42*(8), 916 (1976)

13. Donaldson, E.M.: Talanta, *27*, 79 (1980)
14. Henze, G., Geyer, R., Preuss, I.: Neue Hütte, *14*(7), 438 (1969)
15. Zelikman, A.N., Angelova, V.: Rudodobiv Metal. *24*(3), 42 (1969); Chem. Abstr., *71*, 73010d (1969)
16. Sharipova, N.S., Songina, O.A.: Zh. Anal. Khim., *28*(12), 2348 (1973)
17. Hope. R.P.: Anal. Chem., *29*, 1053 (1957)
18. Budesinsky, B.W.: Analyst (London), *105*, 278 (1980)
19. Pollock, E.N.: Anal. Chim. Acta, *47*, 367 (1969)
20. Dreesen, D.R., Gladney, E.S., Owens, J.W.: J. Water Pollut. Control Fed., *51*, 2447 (1979)
21. Polyakova, V.V., Saraeva, N.F.: Nauchn. Tr. Nauchno-Issled. Inst. Tsvetn. Metal., (48), 70 (1981); Anal. Abstr., *42*, 3B193 (1982)
22. Lontsikh, S.V., Raikhbaum, Ya.D.: Nauchn. Tr. Irkutsk. Gos. Nauchno-Issled. Inst. Redk. Tsvetn. Metal, (17), 74 (1968); Chem. Abstr., *71*, 45468r (1969)
23. Gil'bert, É.N., Androsova, N.V., Badmaeva, Zh.O.: Zh. Anal. Khim., *34*(6), 1150 (1979)
24. Rakhmilevich, N.M.: Zavod. Lab., *30*(4), 507 (1964)
25. Shcherbakov, V.G., Yurkevich, Yu.N., Antonova, R.A.: Sb. Tr. Vses. Nauchno-Issled. Inst. Tverdykh Splavov, (3), 31 (1960); Chem. Abstr., *57*, 1533a (1962)
26. Dekhtrikyan, S.A.: Dokl. Akad. Nauk Arm. SSR, *28*, 213 (1959); Chem. Abstr., *54*, 6402f (1960)
27. Fedorov, M.F., Zotova, G.Ya., Poprukailo, A.M.: Obogashch. Rud., *11*(5), 50 (1966); Chem. Abstr., *66*, 111270v (1967)
28. Minczewski, J., Rozyck, C.: Chem. Anal. (Warsaw), *10*, 965 (1965)
29. Kedrova, Yu.K.: Analiz Rud Tsvetnykh Metal. Produktov Ikh Pererabotki, Sb. Nauch. Trudov, (14), 21 (1958); Chem. Abstr., *53*, 21410h (1959)
30. Kuchmistaya, G.I., Nadezhdina, G.B., Davydova, N.M.: Nauchn. Tr. Nauchno-Issled. Proekt. Inst. Redkomet. Prom-sti, *82*, 58 (1977); Chem. Abstr., *90*, 33524k (1979)
31. Pilipenko, A.T., Shinh, N.M.: Ukr. Khim. Zh., *32*(11), 1211 (1966)
32. Hiskey, C.F., Meloche, V.W.: Ind. Eng. Chem., Anal. Ed., *12*, 503 (1940)
33. Jordanov, N., Pavolova, M., Bojkova, D.: Talanta, *23*, 463 (1976)
34. Nebesar, B.: Anal. Chim. Acta, *39*, 309 (1967)
35. Duca, A., Calu, C.: Analusis, *1*, 365 (1972)
36. Terada, K., Yoshimura, Y., Osaki, S., Kiba, T.: Talanta, *14*, 53 (1967)
37. Zaichikova, L.B., Lutchenko, N.N.: Sb. Nauchn. Tr. Gas. Nauchno-Issled. Inst. Tsvetn. Metal., (18), 45 (1961); Chem. Abstr., *60*, 2324g (1964)
38. Randa, Z., Benada, J., Kuneir, J., Vobecky, M.: Radiochem. Radioanal. Lett., *3*, 227 (1970)
39. Vinogradov, A.V., Dronova, M.I.: Zh. Anal. Khim., *20*(3), 343 (1965)

Chapter 14

Molybdenum and Molybdenum Based Alloys

Pure molybdenum metal is prepared by hydrogen reduction of MoO_3 [1], zone re-
fining [2], electrodeposition from molten salt solution [3], sublimation of $MoCl_5$
followed by hydrogen reduction [4, 5], and in the laboratory by a thermite reduction process [6]. The properties of molybdenum are well known and the more important of these summarized in Table 14-1.

   Molybdenum is an important element for many applications, both in its pure
form and, more extensively, as an alloying element with other metals [11]. Molybdenum and molybdenum alloys are characterized by high temperature strength, for
example, in the manufacture of aircraft engine parts and elsewhere where metal
parts are subjected to high temperatures. Because the coefficient of expansion for
molybdenum is similar to that of some glasses and because of its favorable electri-
cal conductivity, molybdenum finds use in glass metal seals, for example, in the
manufacture of electronic tubes. Numerous other applications of molybdenum exist.

   Properties of molybdenum are greatly effected by the presence of trace impurities in the metal. Their presence can be detrimental. Should, for example, they
increase the brittleness of molybdenum, more rapid part failure could occur. On
the other hand, in the proper combination and concentration they can increase the

Table 14-1. Some Physical Properties of Molybdenum [7–10]

Property	Value
atomic mass	95.94
atomic number	42
boiling point	5560 °C
crystal structure (body-centered cubic)	a = 0.31405 nm (25 °C)
density	10.22 g/cm^3 (20 °C)
electrical resistivity	5.2 μ Ω-cm (20 °C)
heat of fusion	290 joule/g
heat of vaporization	5.12 kjoule/g
melting point	2610 °C
nuclear cross-section (thermal neutrons)	2.5 barn
tensile strength	~65,000 psi
thermal conductivity	1.5 joule/°C-cm-sec
thermal expansion	5.35×10^{-6}/°C
vapor pressure	3.3 μpascal (1600 °C)

strength of molybdenum thus providing improved performance. Gases in particular have a pronounced influence on the physical and mechanical properties of refractory metals, including molybdenum [12, 13]. Nitrogen, oxygen, hydrogen, and carbon, through formation of nitrides, oxides, hydrides, and carbides either with molybdenum or with other metallic impurities in the molybdenum matrix are of great concern to the metallurgist. Especially important is the presence of impurities in interstitial positions along grain boundaries of the matrix material. Also, one must distinguish between surface contamination and contamination within the sample. Failure to correct for surface contamination can greatly effect trace values reported as present in a bulk sample [14]. Vacuum fusion [15], emission spectrometry [16], and activation analysis [17] are but some of the techniques used for analysis of trace gases in molybdenum. These and other methods for determining gases in molybdenum have been reviewed [18] and compaired [19, 20]. Comparison of results for determining oxygen in a molybdenum reference standard by various laboratories is available [21] as are results for determining interstitial components [22]. Auger spectrometry is used to assess the thickness of metal oxide films on a molybdenum base [23]. Auger spectrometry, secondary ion bombardment, and other techniques are all applied to determining trace impurities in molybdenum and molybdenum based alloys [24–26].

Molybdenum is soluble in nitric acid. Molybdenum based alloys may require acid mixtures, hydrogen peroxide, the presence of oxidizing agents, or fusion to place them in solution. Reported procedures include those using mixtures of HNO_3-HCl [27–29], HNO_3-HF [30], and HNO_3-H_3PO_4 [31]. 30% (w/v) hydrogen peroxide [32] and mixtures of H_2O_2–H_2SO_4 [33], H_2O_2–HNO_3 [34], H_2O_2-citric acid [35], and H_2O_2–HCl–HNO_3 [36] also dissolve molybdenum and molybdenum based alloys. Other acid combinations [37–39], fusion [40, 41], and conversion of the metal to its oxide by burning at elevated temperatures [42–45] are possible alternatives in the analysis of molybdenum and molybdenum based alloys.

Donaldson and Inman [37]: Add 20 mL 9 M sulfuric acid, H_2SO_4, 2 mL 15 M nitric acid, HNO_3, and 2 mL 12 M hydrochloric acid, HCl, to 0.5 g powdered sample in a 250 mL beaker. When dissolved, carefully evaporate the solution to fumes of sulfur trioxide, SO_3. Rinse the sides of the beaker with water and again evaporate until SO_3 fumes appear. Dissolve the residue in water or appropriate acid solution.

Spano and Green [38]: A 5 g sample is placed in a 100 mL beaker. Add 15 drops 85% (w/v) phosphoric acid, H_3PO_4 and 5 mL water. Add dropwise a total of 12 mL 15 M nitric acid, HNO_3, heating the sample if necessary at about 80 °C to aid in dissolving. After the sample is dissolved, cool the beaker in a water bath and carefully add 25 mL 30% (w/v) hydrogen peroxide, H_2O_2. Heat the sample on a hot plate until the solution changes from brown, through yellow to a red-brown color. Boil the sample to remove excess H_2O_2 and dilute as necessary before continuing with the determination.

Kallmann, Hobart, Oberthin, and Brienza [39]: Place a 5 g sample in a 200 mL teflon bottle. Add 4 g of potassium chromate-potassium hydroxide mixture. (Prepare mixture by placing 700 g K_2CrO_4 in 1 L water. Add 3 g KOH and stir until the solids dissolve. Evaporate the solution to dryness under a heat lamp. Transfer the residue to a porcelain crucible and place the material in a muffle furnace. Gradually increase the temperature to 500 °C. Remove the crucible. When cool, store the mixture in sealed container.) Add 5 mL 85% (w/v) phosphorus acid, H_3PO_4, and 60 mL 40% (w/v) hydrofluoric acid, HF. Loosely cover the bottle with a teflon square and maintain the sample at a gentle boil on a hot plate. It may be necessary to add additional HF. Approximately 3–4 h is required for the sample to

Table 14-2. Simultaneous Determination of Several Diverse Elements in Pure Molybdenum Metal and Molybdenum Based Alloys

Elements	Concentration Range	Procedure	Ref.
Mo-metal			
Cu, Cu, Fe, Ni, Zn	0.0001–0.001%	colorimetry	[46]
		diacetyldioxime (Ni)	
		β-nitroso-α-naphthol (Co)	
		diethyldithiocarbamate (Cu)	
		1,10-phenanthroline (Fe)	
		anodic stripping for Zn	
Al, Bi, Co, Cr, Cu, Fe, Mg, Ni, Pb, Si, Sn, Ti, V	0.0001–1.0%	emission spectroscopy	[28]
As, Bi, Cd, Cr, Cu, Ni, Pb, Sb, Sn, Ti, Zn	0.0001–0.001%	emission spectroscopy	[47]
Al, Fe, Mg, Si	–	emission spectroscopy	[48]
Co, Cr, Fe, Mn, Ni	0.0005–0.002%	atomic absorption	[35]
Co, Cu, Fe, Mg, Mn, Ni, Pb, Zn	<0.01%	x-ray fluorescence	[38]
Ag, Au, Br, Co, Cu, Nb, Ni, Os, Pd, Pt, Rh, Ru, Zn	0.001–0.01%	x-ray fluorescence	[45]
Mo-based alloys			
Al, B, Cr, Co, Cu, Fe, Mn, Ni, Nb, Si, Ta, Ti, W, V, Zr	–	various methods	[49]

dissolve. After solution is achieved, remove the cover and continue heating to evaporate excess HF. Dilute the sample with water and proceed with the determination.

Püschel and Lassner [40]: Add to a 1 g sample in a platinum crucible 4.0 g of 1:1 (w/w) mixture of sodium carbonate-potassium carbonate, Na_2CO_3–K_2CO_3. Fuse the mixture over a flame for about 15 min. Cool the melt and add a few mL water to the crucible. Place the crucible in a 100 mL beaker. Add 2.0 mL 12 M hydrochloric acid, HCl. Follow this with enough water to dissolve the sample from the crucible. Remove the crucible, rinse it with 2.0 mL additional 12 M HCl followed by water. Collect all rinsings in the 100 mL beaker. Evaporate the sample to a suitable volume, acidify further if necessary, and proceed with the determination.

Table 14-2 lists procedures for simultaneous determination of several impurities in pure molybdenum metal and molybdenum based alloys. Table 14-3 identifies procedures for determination of individual elements in pure molybdenum.

Many alloying elements are combined with molybdenum to produce materials designed for specific purposes. For high temperature high strength applications the molybdenum based alloys TZM (0.5% Ti, 0.08% Zr, 0.015% C) and TZC (1.25% Ti, 0.30% Zr, 0.15% C) are perhaps best known. Analyses of these and other alloys for individual minor and trace components are listed in Table 14-4. Molybdenum-tungsten alloys find application in production of special steels. Their analysis for minor and trace elements (Ti, Zr, Hf, La, and Y) by emission spectrosco-

Table 14-3. Determination of Individual Diverse Elements in Pure Molybdenum Metal

Element	Procedure	Reference
Al	colorimetry 8-hydroxyquinoline	[50]
Al	neutron activation analysis	[51]
As	colorimetry molybdoarsenate	[52]
Bi	emission spectroscopy	[53]
C	combustion collect CO_2	[42, 43]
C	activation analysis	[54, 55]
Co	colorimetry 1-(2-pyridylazo)-2-naphthol	[56]
Cu	colorimetry bathocuproine	[57]
Cu, Cd, Sn	polarography	[58]
Cu, Si	emission spectroscopy	[59]
Fe	colorimetry thiocyanate	[60]
Fe	polarography	[61]
Fe, Mn, Mg	emission spectroscopy	[62]
H_2	manometry	[63]
Mn	colorimetry 1-(2-pyridylazo)-2-naphthol	[37]
N_2	colorimetry indophenol	[39]
N_2	vacuum extraction	[64]
N_2	distillation	[65]
Na, K	activation analysis	[66]
Nb	colorimetry 4-(2-pyridylazo)resorcinol	[67]
Ni	colorimetry 1-(2-pyridylazo)-2-naphthol	[68]
Ni, Co	atomic absorption	[69]
Ni	x-ray fluorescence	[70]
O_2	activation analysis	[71, 72, 73]
O_2	mass spectrometry	[74]
O_2, N_2	gas chromatography	[12]
Pd	atomic absorption	[75]
Pd	neutron activation analysis	[76]
Re	colorimetry hydroxylamine	[77]
Re	neutron activation analysis	[78]
S	conductometry	[79]
Sb	colorimetry methyl violet	[80]

Sb, Bi, Cd, Cu	colorimetry
	    methyl violet (Sb)
	    iodide (Bi)
	    rhodamine B (Cd)
	    diethyldithiocarbamate (Cu)
		[81]

Th	radiochemistry	[82]
Ti	colorimetry hydrogen peroxide	[83]
Ti, Zr	x-ray fluorescence	[84]
W	colorimetry thiocyanate	[85, 86]
W	emission spectroscopy	[87]
W	polarography	[88]
W, Hf	radiochemistry	[89]
Zr	colorimetry xylenol orange	[90]
Zr	emission spectrometry	[91]
Zr, Hf	x-ray fluorescence	[29]
Zr	activation analysis	[92]

Table 14-4. Determination of Individual Diverse Elements in Molybdenum Based Alloys

Element	Procedure	Ref.
B, Zr	emission spectroscopy	[36]
Ca	atomic absorption	[93]
Ca	radiochemistry	[94]
Ce	colorimetry citrate	[30]
La, Y	colorimetry 4-(2-pyridylazo)resorcinol	[95]
Nb	colorimetry 5-ethylamino-2-(2-thiazolylazo)-p-cresol	[41]
Re	colorimetry thiourea	[31]
Re	atomic absorption	[96]
Re	potentiometry	[33, 97]
S	polarography	[32]
Ti	colorimetry diantipyrylmethane	[27]
Ti, Zr	colorimetry tiron (Ti) 8-hydroxyquinoline (Zr)	[98]
Ti, V	emission spectroscopy	[99]
Ti, Zr	emission spectroscopy	[100]
W	gravimetry β-naphthoquinoline	[101]
W	colorimetry thiocyanate	[49]
W	emission spectroscopy	[102]
Y, Sc	colorimetry arsenazo III	[103]
Zr, HF	titrimetry EDTA	[104]
Zr, Ti	colorimetry arsenazo III (Zr) diantipyrylmethane (Ti)	[105]

py [106] and by various individual chemical tests (Al, Fe, Cu, Ni, Mn, P, S, C) [107] appear in the literature.

Sterlace and Wise [28] describe an emission spectroscopic method for determining up to 13 trace impurities in molybdenum metal in the 0.0001 to 1.0% range. Al, Bi, Cr, Co, Cu, Fe, Mg, Ni, Pb, Si, Sn, Ti, and V are measured by this procedure. Following conversion of the sample to molybdenum(VI) oxide, interference from molybdenum itself is reduced by volatilizing the MoO_3 in the presence of GeO_2. A double ignition of the 1:1 (w/w) MoO_3–GeO_2 mixture lowers the molybdenum concentration to less than 0.1%. The authors point out, however, that even at this level some difficulty can be encountered in identifying the characteristic lines of each impurity when molybdenum lines are also present. They recommended that an iron spectrum be obtained under the experimental conditions used for sample and standards as an aid in locating each impurity emission line. Quantitative measurement is obtained by preparing a calibration curve for each impurity based upon standards prepared in a pure molybdenum(VI) oxide matrix and subjected to the same ignition-volatilization as the unknown sample. Following is their procedure.

A standard stock mixture containing 2.000% each of the metals studied is prepared by adding – compound, amount – Al_2O_3 (3.779 g), Bi_2O_3 (2.230 g), Cr_2O_3 (2.923 g), CoO

Operating Conditions

Electrodes	sample (anode)	graphite rod with cavity (Ultra Carbon 320302)
	counter (cathode)	graphite rod (Ultra Carbon 122202)
Excitation	dc arc, 10 amp (Jarrell-Ash Varisource)	
Spectrograph	Jarrell-Ash 3.4 m, Ebert mount, 0.5 nm/mm dispersion in first oder	
Wavelength range	230.0–470.0 nm	
Slit width	25 μ, 7 step filter used at 10% and 100%	
Exposure time	to completion (about 3 min)	
Photographic plates	Kodak spectrum analyzer No. 1	

(2.543 g), CuO (2.504 g), Fe_2O_3 (2.859 g), MgO (3.316 g), NiO (2.545 g), PbO (2.154 g), SiO_2 (4.278 g), SnO_2 (2.539 g), TiO_2 (3.336 g), V_2O_5 (3.570 g), Na_2CO_3 (4.610 g), and ZnO (2.490 g) with sufficient graphite to produce 100.0 g total mixture. Spectroscopically pure reagents are used and the mixture thoroughly ground in a mortor to assure uniformity. Portions of this mixture are diluted with pure molybdenum(VI) oxide to prepare working standards.

Metal Concentration	Amount Stock Mixture, g	Amount MoO_3, g
1.000%	0.500	0.500
0.300%	0.150	0.850
0.100%	0.050	0.950
0.030%	0.150	9.850
0.010%	0.050	9.950
0.003%	0.150	99.85
0.001%	0.050	99.95
0.0003%	0.015	99.98_5
0.0001%	0.005	99.99_5

Samples of presumably pure molybdenum in the form of metal chips are cleaned by rinsing first in acetone and then in distilled water. To remove additional surface contamination they are then soaked in 6 M hydrochloric acid, HCl, for 5 min, removed from the acid rinsed with water, and dried at 140 °C for one hour. Weigh between 0.1 and 0.5 g of dried metal sample and place in a 200 mL tall form beaker. Cover the sample with 10 mL water and 10 mL 15 M nitric acid, HNO_3. Warm the sample on a hot plate adding additional HNO_3 if necessary until it is decomposed into a powdery form. Add 5 mL 12 M HCl and continue heating until the powder dissolves and the solution is clear. Additional HCl and/or HNO_3 may be required. Transfer the clear solution to a 250 mL volumetric flask and add distilled water to a final volume of 250 mL. An aliquot of this solution containing between 30 and 35 mg molybdenum is placed in a Vycor crucible previously ignited at 500 °C to constant weight. Carefully evaporate the water from the crucible. Ignite the dry residue which remains in a furnace at 500–550 °C for 90 min. This converts the molybdenum salts to molybdenum(VI) oxide. Cool and weigh the sample.

Separate weighed portions of each standard and sample are mixed with equal weights of pure germanium(IV) oxide, GeO_2. Use a mortar and pestle to thoroughly grind each standard and/or sample-GeO_2 mixture. Each mixture is transferred to a porcelain crucible and ignited at 700–750 °C for two hours in a furnace. Remove the crucibles, cool, and re-ignite for one additional hour at 700–750 °C. The second ignition is necessary to assure satisfactory volatilization of MoO_3. Using separate electrodes for each standard and sample, fill the crater of each with about 15 mg of material. Place in turn each electrode in the arc stand and excite each until the sample has burned away, about 3 min, using a 10 amp current. Record the spectrum of each standard and sample. Record also a pure iron spectrum to be used later in locating the analytical lines of each impurity element. The spectrum plates are developed, densiometer readings taken, and the data treated by known procedures for emission spectroscopic measurement. Those unfamiliar with these procedures should consult appropriate references [108, 109]. The authors recommends a computer program to aid in handling the calculations. Their program and others are useful for this purpose [110, 111]. Concentrations read from the calibration curves must be corrected for the volume aliquot taken to obtain percent values for each impurity in the original sample.

The following emission lines are monitored for each impurity studied.

element	emission line, nm
Al	308.216
Bi	306.772
Cr	284.325
Co	341.234
Cu	324.754
Fe	282.328
	340.745
Mg	285.213
Ni	341.476
Pb	283.307
Si	251.433
	251.612
Sn	283.999
	317.502
Ti	334.941
V	318.398
	318.540

References

1. Duca, A.: Acad. Repub. Pop. Rom., Fil. Cluj, Stud. Cerce. Stiint., *3*, 42 (1952); Chem. Abstr., *50*, 16596h (1956)
2. Belk, J.A.: J. Less-Common Met., *1*, 50 (1959)
3. Cumings, R.E., Cattoir, F.R., Sullivan, T.A.: U.S. Bur. Mines, Rep. Invest. No. 6850 (1966); Chem. Abstr., *66*, 13110y (1967)
4. Kirshenbaum, N.W., Bakish, R., Badiali, M.: NASA Doc. N 62–10657 (1962); Chem. Abstr., *59*, 14571e (1963)

5. Badiali, M., Bakish, R., Kirshenbaum, N.W.: Reinststoffe Wiss. Tech., Intern. Symp. 1, Dresden 1961, 147 (1963); Chem. Abstr., *60*, 14168h (1964)
6. Braum, H.J.: Metallbörse, *19*, 2190 (1929); Chem. Abstr., *23*, 5425[4] (1929)
7. Killeffer, D.H., Lenz, A.: Molybdenum Compounds, 1952, New York, Wiley-Interscience
8. Molybdenum Metal, 1961, New York, Climax Molybdenum Company
9. Syre, R.: AGARDograph No. 94 (1965)
10. Mann, J.W.: Encyclopedia of Industrial Chemical Analysis Vol. 16 (Snell, F.D., Hilton, C.L. eds.) 1972, New York, Wiley-Interscience
11. Manzone, M.G., Briggs, J.Z.: Molybdenum, Less Common Alloys of Molybdenum, 1962, New York, Climax Molybdenum Company
12. Winge, J.K., Fassel, V.A.: Anal. Chem., *37*, 67 (1965)
13. Hirschfeld, D., Ortner, H.M.: Metall (Berlin), *35*, 434 (1981)
14. Geerts, J., Triffaux, J., Van Audenhove, J.: Comm. Eur. Communities, Rep. EUR No. 6607 (1979); Anal. Abstr., *40*, 4B1 (1981)
15. Klyachko, Yu.A., Izmanova, T.A., Chistyakova, E.M.: Zavod, Lab., *28*, 923 (1963)
16. Malamand, F.: Office Nat. Etud. Rech. Aerosp. (Fr.) Note Tech. No. 105 (1967); Chem. Abstr., *69*, 15865j (1968)
17. Savitskii, E.M., Burkhanov, G.S., Usmanova, M.M., Mukhammedov, S., Kirillova, V.M., Ottenberg, E.V., Mukhamedshina, N.M., Vlasova, I.V.: Zh. Anal. Khim., *36*(8), 1552 (1981)
18. Ortner, H.M.: Talanta, *26*, 629 (1979)
19. Picklo, G.A.: Rep. NRL (Naval Res. Lab. US) Prog. 1 (1961); Chem. Abstr., *62*, 13835c (1965)
20. Vandecasteele, C., Struckman, K., Engelmann, Ch., Ortner, H.M.: Talanta, *28*, 19 (1981)
21. McLauchlan, E.J.: Metall. Met. Form., *43*, 385 (1976)
22. Kallmann, S., Liu, R., Oberthin, H., Stevenson, R., Lux, F.: US Dep. Comm. Rep. AD 622275 (1965); Chem. Abstr., *66*, 91370x (1967)
23. Holloway, D.M.: Appl. Spectrosc., *27*, 95 (1973)
24. Vidal, G.P., Galmand, P., Lanusse, P.: Rech. Aerosp. No. 122, 29 (1968); Chem. Abstr., *69*, 15642j (1968)
25. Ortner, H.M., Lassner, E.: Mikrochim. Acta Suppl., *7*, 41 (1977)
26. Gras, D.J., Limon, A., Waeber, W.B., Brown, J.D.: Mikrochim Acta, I, 85 (1982)
27. Polyak, L.Ya.: Zh. Anal. Khim., *18*(8), 956 (1963)
28. Sterlace, J.S., Wise, W.M.: Anal. Lett., *10*, 769 (1977)
29. Nekhaev, N.N., Trofimov, N.V., Busev, A.I., Petrov, B.I.: Zavod. Lab., *44*(8), 956 (1978)
30. Merkulova, K.S., Bruile, E.S.: Zh. Prikl. Khim. (Leningrad), *46*(3), 655 (1973)
31. Antonova, R.A., Gedevanova, I.V., Suvorova, S.N.: Nauchn. Tr. Vses. Nauchno-Issled. Inst. Proekt. Inst. Tygoplavk. Metal. Tverd. Splavov No. 13, 130 (1973); Chem. Abstr., *80*, 90753b (1974)
32. Kaplan, B.Ya., Shiryaeva, O.A.: Zavod. Lab., *38*(12), 1437 (1972)
33. Zarinskii, V.A., Shpigun, L.K., Trepalina, V.M., Volobueva, I.V.: Zavod. Lab., *43*(8), 941 (1977)
34. Fedoroff, M.: Bull. Soc. Chim. Fr., 3451 (1968)
35. Neumann, G.M.: Fresenius' Z. Anal. Chem., *259*, 337 (1972)
36. Buzasi, A.: Metall (Berlin), *32*, 255 (1978)
37. Donaldson, E.M., Inman, W.R.: Talanta, *13*, 489 (1966)
38. Spano, E.F., Green, T.E.: Anal. Chem., *38*, 1341 (1966)
39. Kallmann, S., Hobart, E.W., Oberthin, H.K., Brienza, W.C.: Anal. Chem., *40*, 332 (1968)

40. Püschel, R., Lassner, E.: J. Less-Common Met., *17*, 313 (1969)
41. Mal'tseva, L.S., Elinson, S.V.: Zavod. Lab., *39*(4), 385 (1973)
42. Aruina, A.S.: Zavod. Lab., *12*, 411 (1946)
43. Fischer, W., Bastius, H.: Metall (Berlin), *14*, 429 (1960)
44. Friedrich, K., Engelhardt, H.: Fresenius' Z. Anal. Chem., *249*, 244 (1970)
45. Markov, V.L., Shvangiradze, R.R.: Zh. Anal. Khim., *33*(1), 107 (1978)
46. Grossmann, O., Döge, H.G., Grosse-Ruyken, H.: Fresenius' Z. Anal. Chem., *219*, 48 (1966)
47. Britshe, M.E., Bronshteïn, A.N., Gerken, E.B., Ivantsov, L.M.: Mat. Vses. Sovesh. Spektrok.-Anal. Tsvet., 9 (1955); Chem. Abstr., *54*, 5335e (1960)
48. Shcherbakov, V.G., Anikeeva, N.P., Ignatova, A.Ya., Magala, T.Z.: Sb. Tr. Vses. Nauchno-Issled. Inst. Tverdykh Splavov, (3), 56 (1960); Chem. Abstr., *57*, 4009b (1962)
49. Bush, G.H., Higgs, D.G.: Analyst (London), *80*, 536 (1955)
50. Gorlova, M.N.: Zavod. Lab., *40*(1), 27 (1974)
51. Fedoroff, M.: C.R. Acad. Sci. Paris, Ser. C., *267*, 1227 (1968)
52. Shcherbakov, V.G., Onuchina, G.V.: Khim. Svoistva Metody Anal. Tugoplavkikh Soedin., 128 (1969); Chem. Abstr., *72*, 38513u (1970)
53. Muntz, J.H.: Dev. Appl. Spectrosc., *3*, 298 (1963)
54. Gorenko, A.F., Skakun, N.A., Zadvornyi, A.S., Bugaeva, N.I., Shevchenko, G.M., Klyucharev, A.P.: Zh. Anal. Khim., *28*(6), 1227 (1973)
55. Rodionov, V.I., Samosyuk, V.N., Chapyzhnikov, B.A., Revel, G., Fedoroff, M.: Radiochem. Radioanal. Lett., *18*, 379 (1974)
56. Püschel, R., Lassner, E., Illaszewicz, A.: Chem.-Anal., *52*, 40 (1966)
57. Penner, E.M., Inman, W.R.: Talanta, *10*, 407 (1963)
58. Yakovlev, P.Ya., Dymova, M.S.: Sb. Tr. Tsentr. Nauchno-Issled. Inst. Chern. Metall. No. 24, 133 (1962); Chem. Abstr., *59*, 4533f (1963)
59. Shaevich, A.B., Perepelkina, M.A., Korovina, A.G.: Byul. Nauchn.-Tekhn. Inform. Ural. Nauchno-Issled. Inst. Chern. Metall. No. 8, 111 (1960); Chem. Abstr., *57*, 4020a (1962)
60. Shcherbakov, V.G.: Sb. Tr. Vses. Nauchno-Issled. Inst. Tverd. Splavov No. 5, 303 (1964); Chem. Abstr., *62*, 1071f (1965)
61. Fukker, K., Hegedus, A.J.: Mikrochim. Acta, 92 (1961)
62. Nakashima, R., Sasaki, S.: Anal. Chim. Acta, *85*, 75 (1976)
63. Petushkov, E.E., Tserfas, A.A., Maksumov, T.M.: Metody Opred. Issled. Sostoyaniya Gaz. Metall., 266 (1968); Chem. Abstr., *71*, 27186t (1969)
64. Zav'yzlov, O.V., Karpov, Yu.A.: Zh. Anal. Khim., *29*(8), 1620 (1974)
65. Grallath, E., Ortner, H.M.: Talanta, *25*, 195 (1978)
66. Döge, H.G.: Anal. Chim. Acta, *38*, 207 (1967)
67. Wood, D.F., Jones, J.T.: Analyst (London), *93*, 131 (1968)
68. Püschel, R., Lassner, E.: Mikrochim. Acta, 17 (1965)
69. Merzlyakov, A.V., Malinina, R.D., Solomatin, V.T., Nuzhdina, V.N.: Zavod. Lab., *42*(11), 1331 (1976)
70. Bibinov, S.A., Kim, Yu.S., Mukhamedshina, N.M., Shegai, A.A.: Zavod. Lab., *42*(8), 948 (1976)
71. Andreev, A.V., Barit, I.Ya., Musaelyan, R.M., Pronman, I.M.: Zh. Anal. Khim., *21*(12), 1453 (1966)
72. Abdurakhmanova, S.R., Kireev, V.A., Navalikhin, L.V., Talanin, Yu.N.: Zh. Anal. Khim., *23*(8), 1188 (1968)
73. Faure-Mazagol, L., Tousset, J., Boissier, M.: Analusis, *2*, 287 (1973)
74. Glinskikh, V.M., Melashvili, V.A., Ordzhonikidze, K.G., Samadashvili, O.A.: Zh. Anal. Khim., *29*(1), 116 (1974)

75. Begak, O.Yu., Nikolaev, G.I., Pokrovskaya, K.A.: Zavod. Lab., *40*(8), 963 (1974)
76. Fedoroff, M.: C.R. Acad. Sci. Paris, Ser. C, *270*, 486 (1970)
77. Borisova, L.V., Ermakov, A.N., Ismagulova, A.B.: Analyst (London), *107*, 495 (1982)
78. Samadi, A.A., Ailloud, P., Fedoroff, M.: Anal. Chem., *47*, 1847 (1975)
79. Yurkevich, Yu.N., Shcherbakov, V.G.: Zh. Anal. Khim., *16*, 617 (1961)
80. Lazarev, A.I., Lazareva, V.I.: Zavod. Lab., *25*, 400 (1959)
81. Ryabchikov, D.I., Lazarev, A.I.: Renii, Tr. Vses. Sovesh. Probl. Reniya, Akad. Nauk SSSR, Inst. Metall., 267 (1958); Chem. Abstr., *57*, 1524e (1962)
82. Aleksandrov, L.N.: Zavod Lab., *26*, 975 (1960)
83. Čížek, Z., Doležal, J., Sulcek, Z.: Anal. Chim. Acta, *100*, 479 (1978)
84. Wurzinger, H., Müller, K.: Fresenius' Z. Anal. Chem., *284*, 101 (1977)
85. Vinogradov, A.V., Dronova, M.I.: Zh. Anal. Khim., *20*, 343 (1965)
86. Reef, B., Döge, H.G.: Talanta, *14*, 967 (1967)
87. Tarasevich, N.I., Khlystova, A.D., Pak, E.A.: Zavod. Lab., *25*, 955 (1959)
88. Herion, G., Scholz, F., Andreas, B.: Z. Chem., *19*(3), 116 (1979)
89. Leushkina, G.V., Lobanov, E.M., Dutov, A.G., Matveeva, N.P.: Zavod. Lab., *35*(6), 715 (1969)
90. Elinson, S.V., Nezhnova, T.I.: Zavod. Lab., *30*(4), 396 (1964)
91. Frishberg, A.A.: Zh. Prikl. Spektrosk., *5*, 12 (1966); Chem. Abstr., *65*, 19309c (1966)
92. Oblivantsev, A.N., Yakovlev, B.M.: Tr. Nauchno-Issled. Inst. Yader. Fiz. Elektron. Avtom. Tomsk. Politekhn. Inst. No. 6, 26 (1976); Chem. Abstr., *86*, 37259t (1977)
93. Zhukova, M.P., Solomatin, V.T., Syshchuk, L.A.: Tekhnol. Legk. Splavov. Nauchn-Tekhn. Byul. VILSA No. 3, 42 (1977); Chem. Abstr., *88*, 98653e (1978)
94. Zhukova, M.P., Shemyakin, F.M., Yakovlev, P.Ya.: Zh. Anal. Khim., *28*(9), 1705 (1973)
95. Kitaeva, L.P., Volynets, M.P., Makulov, N.A., Surorova, S.N.: Zh. Anal. Khim., *36*(1), 102 (1981)
96. Saltykova, A.M., Blokhina, D.I.: Nauchn. Tr. Nauchno-Issled. Proekt. Inst. Red-komet. Prom-sti No. 82, 90 (1977); Chem. Abstr., *90*, 15767n (1979)
97. Pavlova, M., Kadieva, S., Jordanov, N.: Fresenius' Z. Anal. Chem., *285*, 271 (1977)
98. Shakhashiro, M., Freund, H.: Anal. Chim. Acta, *33*, 597 (1965)
99. Raff, E.L.: Zavod. Lab., *31*(2), 184 (1965)
100. Muntz, J.H.: US At. Energy Comm. ML-TDR-64-36 (1964); Chem. Abstr., *62*, 2237g (1965)
101. Mukhina, Z.S., Il'ina, L.I., Kondukova, N.S.: Khim. Svoistva Metody Anal. Tugo-plavkikh Soedin, 108 (1969); Chem. Abstr., *72*, 50661y (1970)
102. Khlystova, A.D., Kabesheva, E.T.: Vestn. Mosk. Univ., Khim., *12*, 98 (1971); Chem. Abstr., *74*, 134584e (1971)
103. Spitsyn, P.K.: Zh. Anal. Khim., *36*(7), 1330 (1981)
104. Ryabchikov, D.I., Bukhtiarov, V.E.: Zh. Anal. Khim., *19*(11), 1411 (1964)
105. Polyak, L.Ya.: Zavod. Lab., *32*(11), 1317 (1966)
106. Lenskaya, K.K., Tikhomirova, O.F., Golubeva, V.N., Sorokina, N.N., Suchelenkova, L.M.: Sb. Tr. Tsentr. Nauchno-Issled. Inst. Chern. Metall. No. 49, 48 (1966); Chem. Abstr., *66*, 25846f (1967)
107. Kral, S.: Hutn. Listy, *22*, 199 (1967)
108. Brode, W.R.: Chemical Spectroscopy 1943 New York, Wiley & Sons
109. Anderson, J.W.: Applied Atomic Spectroscopy Vol. 1 (Grove, E.L. ed.) 1978, New York, Plenum Press
110. Strauss, B.H.: Talanta, *24*, 524 (1977)
111. Matherny, M., Ondáš, J.: Anal. Chim. Acta, *133*, 51 (1981)

Chapter 15

Molybdenum in Ferrous Alloys

Molybdenum is an important alloying element in high temperature high strength steels. It also appears in other types of steel and in various iron based alloys either alone or in combination with other alloying elements [1, 2]. Many common alloyed steels contain a few tenths of a percent molybdenum. Its concentration can range up to several percent in certain tool steels. Nearly every technique known for determining molybdenum has been applied to iron and steel analysis and every company engaged in production of goods containing iron and steel has its own preferred procedure for analyzing its products. In addition to official methods [3] other compilations pertaining to the complete analysis of iron and steel are available [4–6]. Gravimetric and titrimetric methods for molybdenum in steels have been reviewed [7]. Separation of molybdenum from other constituents in iron alloys is discussed in chapter 3.

Ferromolybdenum, containing upwards of 50% molybdenum, is a common source of molybdenum for iron and steel industries. Knowledge of its trace element content is essential prior to its use in manufacture. Ferromolybdenum samples are generally soluble in nitric acid [8] or if an insoluble residue remains it can be treated with hydrofluoric acid [9] or fused with potassium pyrosulfate, $K_2S_2O_7$ [10]. Sodium peroxide, Na_2O_2, fusion also renders ferromolybdenum soluble [11]. Nonmetallic inclusions in ferromolybdenum have been studied by electron probe microanalysis [12] and other techniques [13]. Table 15-1 lists procedures for determining trace impurities in ferromolybdenum.

Silverman [8] – *ferromolybdenum*: A 1 g sample is placed in a tall form 300 mL beaker. Add 20 mL 5 M nitric acid, HNO_3, and heat as necessary to dissolve the majority of the sample. Add 10 mL 18 M sulfuric acid, H_2SO_4, and evaporate the solution until fumes of sulfur trioxide, SO_3, appear. Add to the cooled residue 75 mL 2.5 M hydrochloric acid, HCl. Heat until all soluble salts are dissolved and filter the solution through a medium filter paper (e.g. Whatman number 40). Wash the precipitate with 20–30 mL 2.5 M HCl and then with 20–30 mL distilled water. Repeat the HCl-water washings twice more. If iron salts still remain, yellow color, continue washing with hot 2.5 M HCl and hot water until they are removed. Collect the filtrate and washings, dilute to a known volume, and proceed with the determination of molybdenum and/or other soluble constituents.

Ashy and Headridge [29] – *ferromolybdenum*: In a teflon beaker place an 0.6 g sample. Cover with 5 mL 6 M sulfuric acid, H_2SO_4. Add slowly 3 mL 15 M nitric acid, HNO_3, followed by 2 mL 40% (w/v) hydrofluoric acid, HF. Heat the sample on a hot plate and when dissolved add an additional 4 mL 18 M H_2SO_4. Carefully evaporate the solution until fumes of sulfur trioxide, SO_3, appear. Continue fuming the sample for 10 min. This should be done in a *well ventilated fume hood* to avoid contact with SO_3. Cool the residue and add 15 mL distilled water. Warm the sample on a hot plate to aid in dissolving the residue. Dilute the sample to a desired volume and proceed with the determination.

Table 15.1. Determination of Diverse Elements in Ferromolybdenum

Element	Procedure	Ref.
Al, S	gravimetry	[14]
	8-hydroxyquinoline (Al)	
	barium sulfate (S)	
As	titrimetry iodine	[15]
As, Bi, Cd, Pb, Sb, Sn, Zn	emission spectroscopy	[16, 17]
As, Bi, Pb, Sb, Sn	emission spectroscopy	[18]
Bi	colorimetry iodide	[19]
Cr	colorimetry diphenylcarbazide	[20]
Cu	gravimetry α-benzoinoxime	[8]
Cu	titrimetry ascorbic acid	[21]
Fe, Si	x-ray fluorescence	[22]
Mn	polarography	[23]
P	gravimetry magnesium ammonium phosphate	[9]
P	colorimetry reduced molybdophosphate	[24]
Sb	colorimetry methyl violet	[25]
Si	colorimetry reduced molybdosilicate	[26]
Sn	gravimetry sulfide	[27]
Sn	colorimetry phenylfluorene	[28]
Ti	colorimetry alizarinsulfonate	[10]
W	colorimetry thiocyanate	[11]

The association of molybdenum with carbon, boron, and silicon in steels, as molybdenum carbide, boride, or silicide, is of interest in regulating the properties of a particular steel. Separation of these compounds by electrolysis [30, 31] or other means [32, 33] and their quantitative and structural characteristics [34–36] have been studied. Oxide formation and its effect on the properties of a particular steel are also available [37, 38].

Iron and steel samples are generally dissolved in nitric acid although other acids and combinations of acids are used. Frequently complexing agents are added as an aid in dissolving otherwise insoluble constituents or to sequester those components that would otherwise interfere with determination of the desired constituent. Reducing agents are sometimes used to reduce possible interfering ions to lower, non-interfering, oxidation states.

Freegarde and Jones [39] – *low alloy steels*: One gram of steel sample is dissolved in 20 mL 15 M nitric acid, HNO_3. When reaction subsides, add 3 mL 15 M phosphoric acid, H_3PO_4. Evaporate the sample to a small volume and add 10 mL 15 M HNO_3. Repeat the evaporation. If carbide residues still persist, add additional HNO_3/H_3PO_4 until they are completely dissolved. Add 20% (w/v) hydrogen peroxide, H_2O_2, dropwise to reduce manganese, if present, and then boil the sample to remove the excess H_2O_2. Filter the sample, if necessary, through a coarse filter paper (e.g. Whatman number 541) to remove undissolved silica, cool, and dilute the filtrate to a known volume.

Purushottam, Naidu, and Lal [40] – *various steels*: Place 0.1–0.5 g sample in a stoppered 25 mL graduated cylinder. Add 5 mL 12 M hydrochloric acid – 15 M nitric acid mixture (HCl: HNO_3, 3:1 v/v) and digest the sample for one hour on a hot plate. After the sample has dissolved, add 0.5 mL 15 M phosphoric acid, H_3PO_4, and dilute to 25 mL with water. Details for the continuation of this procedure for determination of molybdenum by atomic absorption spectrometry are given in chapter 8.

Braithwaite and Hobson [41] – *tungsten steel*: Add 15 mL 12 M hydrochloric acid, HCl, to a 0.5 g sample in a beaker. Heat the sample until the majority of it dissolves and the solution is clear. Carefully add dropwise 3 mL 15 M nitric acid, HNO_3. When reaction has ceased, add 10 mL 9 M sulfuric acid, H_2SO_4, and evaporate the sample until fumes of sulfur trioxide, SO_3, appear. Continue the fuming for 2 min after which time the sample is cooled to ambient temperature. Add 30 mL distilled water and boil the sample to re-dissolve the residue. Cool and add 10 mL 50% (w/v) ammonium citrate, $(NH_4)_2HC_6H_5O_7$. Add one mL saturated sulfur dioxide, SO_2, solution. After mixing add slowly 15 mL 15 M ammonia, NH_3, solution until the sample is neutral to litmus. Add 3–4 mL excess NH_3 solution. Heat, if necessary, to dissolve the insoluble tungsten residue. When the solution is clear, neutralize the sample with 9 M H_2SO_4 and then add 25 mL 9 M H_2SO_4 in excess. Dilute the sample to a known volume with distilled water and proceed with the determination of molybdenum.

Castillo, Belarra, and Aznarez [42] describe a rapid atomic absorption procedure for determination of molybdenum in cast iron, nickel, nickel-chromium, and stainless steels. Molybdenum is first separated from the sample by extraction as its α-benzoinoxime complex and the extract aspirated directly into a nitrous oxide-acetylene flame. A plot of absorbance versus amount of molybdenum is linear up to 10 µg Mo/mL in the aqueous phase prior to extraction. Sensitivity of the method, 0.0044 absorbance reading, is 0.08 µg Mo/mL in the aqueous phase. Molybdenum content of the test samples ranged from 0.08% to 0.43%. Tungsten in a 100 fold excess over molybdenum did not interfere nor did 2 mg Fe/mL in the aqueous sample phase. Readings were made upon the 313.3 nm molybdenum line. Other parameters include spectral bandpass 0.08 nm, 10 mA lamp current, and 2.2 mL/min aspiration rate.

A *standard molybdenum solution* is prepared by dissolving 0.184 g ammonium molybdate, $(NH_4)_6Mo_7O_{24}\cdot4H_2O$ in 1% (v/v) perchloric acid. (Prepare by diluting 10 mL 70% (w/v) $HClO_4$ to one liter with double-distilled water.) Dilute the sample to one liter in a volumetric flask with additional 1% (v/v) $HClO_4$. Exactly 10.0 mL of this solution is tranferred to an empty 100 mL volumetric flask and additional 1% (v/v) $HClO_4$ added until the final volume is 100 mL. This diluted stock solution contains 10.0 µg Mo/mL.

Between 0.1 and 1 g sample, accurately weighed and containing no more than 25 mg Mo, is placed in a 400 mL beaker and covered with 10 mL 6 M hydrochloric acid, HCl. Add several drops 15 M nitric acid, HNO_3. Cover the beaker with a watch glass and heat the sample gently on a hot plate until the sample dissolves. It may be necessary to add additional amounts of the acids. Boil the sample to remove traces of nitrogen oxides. Place the uncovered beaker in a boiling water bath and evaporate the sample to near dryness. Add 20 mL 1% (v/v) $HClO_4$ to dissolve the residue. If a white solid remains, filter the sample through a medium filter paper (e.g. Whatman number 40). Wash the residue thoroughly with 1% (v/v) $HClO_4$ and collect filtrate and washings in a 250 mL volumetric flask. Add additional 1% (v/v) $HClO_4$ to the flask until the final volume is 250 mL.

To a series of 125 mL separatory funnels place 0.0, 1.0, 2.0, 5.0, 7.0, and 10.0 mL diluted molybdenum stock solution. This corresponds, respectively, to 0.0, 10, 20, 50, 70, and 100 µg Mo in each funnel. Add 1% (v/v) $HClO_4$ to those funnels containing less than 10 mL total volume until each volume is ten mL. Place 10.0 mL of the sample solution into a 125 mL separatory funnel. Add to each funnel 3 mL 2% (w/v) α-benzoinoxime, $C_{14}H_{13}NO_2$, (dissolved in ethanol). Mix the contents of each flask and set it aside for 5 min to assure complete reaction between molybdenum and the reagent. Add to each flask 10 mL methyl isobutyl ketone, $C_6H_{12}O$. Shake each flask for one minute and then wait several minutes for the layers to separate. Transfer each organic phase to a 10 mL volumetric flask and adjust the volume of each to exactly 10 mL using additional methyl isobutyl ketone. Acetylene flow is approximately 3.8 L/min and nitrous oxide flow approximately 5.0 L/min. Exact flow and burner height should be finely adjusted to give maximum absorbance reading for each particular spectrometer using these suggested settings as starting points. Once optimum settings are achieved, aspirate in turn the blank, each standard solution, and the unknown sample into the flame. Record the absorbance readings. Correct, by subtraction, all absorbance readings for the blank reading. Prepare a calibration curve of absorbance reading versus amount of molybdenum, mg, persent in each standard. From the curve read the amount of molybdenum present in the unknown sample. Correct this amount for dilution made during sample preparation and report percent molybdenum in the original sample.

References

1. Gregg, J.L.: The Alloys of Iron and Molybdenum, 1932, New York. McGraw-Hill
2. Molybdenum in Steels, 1941, New York, Climax Molybdenum Company
3. 1979 Annual Book of ASTM Standards, Part 12: Chemical Analysis of Metals; Sampling and Analysis of Metal Bearing Ores, 1979, Philadelphia, American Society for Testing and Materials
4. Lundell, G.E.F., Hoffman, J.I., Bright, H.A.: Chemical Analysis of Iron and Steel, 1931, New York, Wiley
5. Welcher, F.J.: Standard Methods of Chemical Analysis, 6th ed., Vol. II, 1963, Princeton, New Jersey, D. van Nostrand Company
6. Harrison, T.S.: Handbook of Analytical Control of Iron and Steel Production, 1979, Chichester, Horwood
7. Klinger, P.: Arch. Eisenhüttenwes., *14*, 157 (1940)
8. Silverman, L.: Ind. Eng. Chem., Anal. Ed., *12*, 343 (1940)
9. Silverman, L.: Ind. Eng. Chem., Anal. Ed., *13*, 602 (1941)
10. Goto, H., Kakita, Y., Hirokawa, K.: Nippon Kagaku Zasshi, *78*, 1343 (1957); Chem. Abstr., *52*, 13536e (1958)
11. Kalinskii, Ya.M., Naruta, E.G., Chabanenko, N.I., Nikiforova, K.I.: Sb. Tr. Chelyabinsk. Elektromet. Komb. No. 1, 182 (1968); Chem. Abstr., *71*, 56358f (1969)
12. Nicodemi, W., Zoja, R., Caloni, O.: Ric. Sci., *36*, 1261 (1966); Chem. Abstr., *66*, 107187f (1967)
13. Medvedeva, G.A.: Method. Anal. Chern. Tsvet. Metall., 65 (1953); Chem. Abstr., *51*, 1774f (1957)
14. Kakita, Y., Sudo, E., Namiki, M.: Sci. Rep. Res. Inst. Tohoku Univ., Ser. A., *13*, 199 (1961); Chem. Abstr., *56*, 4082i (1962)
15. Binder, O.: Chem.-Ztg., *42*, 619 (1918)
16. Shubina, S.B., Shaevich, A.B., Basova, E.P.: Zavod. Lab., *26*, 1364 (1960)
17. Deliiska, A., Pruvcheva, Kh.: Metalurgia (Sofia), *30*(2), 19 (1975); Chem. Abstr., *83*, 90339h (1975)
18. Tumanova, T.G.: Sb. Tr. Chelyabinsk. Elektromet. Komb, No. 1, 185 (1968); Chem. Abstr., *71*, 56266z (1969)
19. Carlstrom, C.G., Palvarinne, V.: Jernkontorets Ann., *146*, 453 (1962); Chem. Abstr., *57*, 15790h (1962)
20. Imai, T., Nagumo, S.: Kogyo Kagaku Zasshi, *61*, 53 (1958); Chem. Abstr., *53*, 16814i (1959)

21. Khanna, V.B., Sood, B.M.: ISI (Indian Stand. Inst.) Bull., *21*, 133 (1969)
22. Tsukamoto, A., Shimizu, I.: Nippon Kinzoku Gakkaishi, *32*, 473 (1968); Chem. Abstr., *69*, 64307n (1968)
23. Degterev, N.M.: Zavod. Lab., *21*, 917 (1955)
24. Maekawa, S., Ebihara, M.: Bunseki Kagaku, *9*, 731 (1960)
25. Silaeva, E.V., Kurbatova, V.I.: Zavod. Lab., *28*, 280 (1962)
26. Kurbatova, V.I., Silaeva, E.V.: Tr. Vses. Nauchno-Issled. Inst. Standartn. Obraztsov Spektrosk. Etalenov No.1, 79 (1964); Chem. Abstr., *64*, 1347b (1966)
27. Shkotova, S.N.: Zavod. Lab., *6*, 745 (1937)
28. Silaeva, E.V., Kurbatova, V.I.: Zavod. Lab., *27*, 1462 (1961)
29. Ashy, M.A., Headridge, J.B.: Anal. Chim. Acta, *59*, 217 (1972)
30. Popova, N.M., Platonova, A.F.: Zavod. Lab., *23*, 269 (1957)
31. Sosulina, I.A., Varakina, L.P.: Zavod. Lab., *42*, 22 (1976)
32. Kriege, O.H.: US At. Energy Comm. LA-2306 (1959); Chem. Abstr., *55*, 7140e (1961)
33. Opravil, O., Kačerová, O., Pažitný, J., Svatik, I.: Hutn. Listy, *15*, 628 (1960); Chem. Abstr., *55*, 17351i (1961)
34. Kugai, L.N., Nazarchuk, T.N.: Zh. Anal. Khim., *17*, 1082 (1962)
35. Tunney, R.J., Lorimer, G.W., Ridley, N.: Met. Sci., *12*, 271 (1978); Chem. Abstr., *89*, 201511d (1978)
36. Sato, T., Nishizawa, T., Tamaki, K.: Trans. Jpn. Inst. Met., *3*, 196 (1962); Chem. Abstr., *59*, 2444d (1963)
37. Winkler-Gniewek, W., Zechmeister, H., Zeilinger, H.: Mikrochim. Acta Suppl., *8*, 159 (1979)
38. Tokareva, L.B., Voropaeva, L.I.: Khim. Prom-st., Ser.: Reakt Osobo Chist. Veshchest-va No.5, 28 (1979); Chem. Abstr., *93*, 87882f (1980)
39. Freegarde, M., Jones, B.: Analyst (London), *84*, 393 (1959)
40. Purushottam, A., Naidu, P.P., Lal, S.S.: Talanta, *19*, 1193 (1972)
41. Braithwaite, K., Hobson, J.D.: Analyst (London), *93*, 633 (1968)
42. Castillo, J.R., Belarra, M.A., Aznarez, J.: At. Spectrosc., *3*, 58 (1982)

Chapter 16

Molybdenum in Non-Ferrous Alloys

Molybdenum appears as a compound in numerous non-ferrous alloys. Added generally to increase high temperature strength and corrosion resistance, its use for these purposes is increasing as present day technology finds additional applications for alloys with these properties. Aircraft engine parts, nuclear reactor heat exchanges, sampling containers for stack emissions, and equipment for the manufacture of certain chemicals are but some of the uses demanding high temperature, high strength, corrosion resistant alloys. Table 16-1 lists some of the principal nonferrous alloys to which small amounts of molybdenum are added. Molybdenum is capable of alloying with many other elements and, no doubt, additional applications of molybdenum in new and different alloys will develop in the future [1].

Methods for dissolving non-ferrous alloys vary depending upon the particular matrix element. Once dissolved, procedures for molybdenum and other constituents vary depending upon possible interfering reactions between the components present. Procedures for determining molybdenum in non-ferrous alloys are cited in most of the earlier chapters of this book describing various analytical techniques.

Saltykova, Davidovich, and Melamed [5] – *Nb alloys*: In a small beaker place 0.1 g sample, 1.5 mL 18 M sulfuric acid, H_2SO_4, and 0.3 g ammonium sulfate, $(NH_4)_2SO_4$. Mix the contents of the beaker and heat to dissolve the sample. Quantitatively transfer the solution to a 50 mL volumetric flask. Add 2 mL 30% (w/v) hydrogen peroxide, H_2O_2. Add sufficient distilled water to make a final volume of 50 mL.

Table 16.1. Some Desirable Properties of Non-Ferrous Molybdenum Containing Alloys [1, 2, 3, 4]

Matrix Element	Property
Al	corrosion and heat resistance
Co	high temperature strength, corrosion resistance
Cu	long wear friction bearings
Nb	high temperature strength, corrosion resistance
Ni	high temperature strength, corrosion resistance
Ti	high strength to weight ratio, corrosion and heat resistance
U	nuclear fuel elements
W	high temperature strength
Zr	corrosion and heat resistance

Montgomery [6] – *Ni alloy*: A 1 g sample is placed in a teflon beaker. Add 10 mL hydrochloric acid-nitric acid mixture. (Prepare by mixing together 30 mL 12 M HCl and 10 mL 15 M HNO_3.) Cover the beaker and allow the decomposition to proceed. When dissolution is complete, remove the lid and carefully evaporate the sample to near dryness. Dissolve the residue in 5 mL 11 M HCl and 15 mL 40% (w/v) hydrofluoric acid, HF. Again, carefully evaporate the sample to near dryness. Repeat the addition of 5 mL 11 M HCl plus 15 mL 40% (w/v) HF. For a third time evaporate the sample to near dryness but avoid any prolonged baking of the sample. Add 5 mL 11 M HCl and 15 mL 40% (w/v) HF. Transfer the dissolved sample to a 250 mL volumetric flask using distilled water to thoroughly rinse the beaker. Dilute the sample to 250 mL with additional distilled water.

Codell, Mikula, and Norwitz [7] – *Ti alloy*: One gram of sample is placed in a 250 mL beaker and covered wth 30 mL 12 M hydrochloric acid. Heat on a hot plate, if necessary, to achieve dissolution. Add 1 mL 15 M nitric acid, HNO_3, and 5 mL 18 M sulfuric acid, H_2SO_4. Evaporate the sample to fumes of sulfur trioxide, SO_3. Add 10 mL distilled water to the cooled residue to dissolve the sample. Add 20 mL 12 M HCl. Transfer the sample quantitatively to a 50 mL volumetric flask and add sufficient distilled water until the final volume is 50 mL.

Scarborough [8] – *U alloy*: A 4 g sample of alloy is placed in a 400 mL teflon beaker and covered with 20 mL distilled water. Add 60 mL hydrochloric acid-nitric acid mixture. (Prepare by mixing 50 mL 12 M HCl and 10 mL 15 M HNO_3.) Follow this immediately with 10 drops 40% (w/v) hydrofluoric acid, HF. Cover the beaker. Swirl the sample to aid in dissolving. When dissolved, add sufficient distilled water until the final volume is 100 mL. Store the sample in a polyethylene bottle until ready for use.

Pashchenko, Mal'tsev, and Zherikova [9] – *W alloy*: In a 150 mL beaker place 0.1 g sample. Cover this with 3 mL 18 M sulfuric acid, H_2SO_4. Add 2 g ammonium sulfate, $(NH_4)_2SO_4$. Stir the mixture and heat on a hot plate. When activity within the beaker ceases, cool the sample. Add 30 mL distilled water and 30 mL 20% (w/v) sodium hydroxide, NaOH. Stir the solution and add 40 mL additional distilled water. Mix thoroughly and filter any residue that remains through a medium filter paper (e.g. Whatman number 40). Wash the residue on the paper with 10 mL 2% (w/v) NaOH. Collect filtrate and washings in a 250 mL beaker. Add slowly 1 M H_2SO_4 until the sample is pH 9 (phenolphthalein transition from red to colorless). Transfer the sample to a 250 mL volumetric flask, acidify as necessary for the desired procedure to be used in determining molybdenum, and dilute the solution to 250 mL with distilled water.

Elinson, Savvin, and Nezhnova [10] – *Zr alloy*: Place a 1 g sample in a 100 mL beaker. Add 1 g ammonium sulfate, $(NH_4)_2SO_4$, and 1.5 mL 18 M sulfuric acid, H_2SO_4. Heat the sample until it dissolves. Add 2–3 drops 15 M nitric acid, HNO_3, and continue heating the solution until fumes of sulfur trioxide, SO_3, are observed. Add an additional 2–3 drops 15 M HNO_3 and again evaporate the sample to fumes of SO_3. When cool, transfer the residue to a 100 mL volumetric flask using distilled water to wash the sample beaker. Fill the flask to exactly 100 mL with additional distilled water.

Rigdon and Harrar [11] determine molybdenum in a *nuclear reactor alloy* of molybdenum-tungsten-rhenium, containing approximately 18% Mo, by controlled potential coulometry of Mo(VI) to Mo(V) in an oxalate-sulfuric acid medium at pH 2.1. Under their conditions tungsten interference is negligible and rhenium is removed by volatilization prior to molybdenum reduction. Trace amounts of Bi(III), Cu(II), Fe(III), and U(VI) do interfere but corrections can be made for their presence. Consult the original article for details of these corrections. Moderate amounts of Cl^-, ClO_4^-, NO_3^-, and PO_4^{3-} do not interfere in oxalate medium. For solutions containing approximately one mg Mo, relative standard deviation of repetitive measurements was 0.20%. For samples containing approximate-

Operating Conditions

Instrumentation	potentiostat, preferably with integrator (Tacussel model PRT 20-010 MOD or equivalent [12])
Electrodes	working electrode
	Pt foil
	counter electrode
	mercury pool
	reference electrode
	saturated calomel
Cell	20 mL capacity
Temperature	ambient
Decomposition potential	-0.25 V vs. SCE
Deaeration	N_2 for 7 min prior to measurement

ly ten mg Mo, the corresponding deviation was 0.06%. Following are details of their procedure.

Between 0.1 and 1.0 g of sample, accurately weighed, is placed in a 400 mL teflon beaker and covered with 20 mL distilled water. Add to the beaker 5 mL each 18 M sulfuric acid, H_2SO_4, 40% (w/v) hydrofluoric acid, HF, and 15 M nitric acid, HNO_3. Cover the beaker and heat it gently until the sample is dissolved. Remove the cover, rinse the sides of the beaker with water, and carefully evaporate the solution until dense fumes of sulfur trioxide, SO_3, appear. Cool the residue and quantitatively transfer it to a 500 mL tall form beaker. This large size beaker is helpful in preventing molybdenum loss during the following steps. Cover the beaker with a watch glass supported upon glass hooks to raise it above the rim of the beaker. Evaporate the solution until fumes of SO_3 appear. Increase the temperature of the sample thereby generating copious fumes of SO_3. Continue fuming until only one or two mL of H_2SO_4 remain in the beaker. This is necessary to assure volatilization of rhenium present in the samples and should be carried out in a well ventilated fume hood. Cool the moist residue and add 10 mL 6 M sodium hydroxide, or more, to dissolve the solids present. Adjust the sample to pH 11 by adding additional 6 M NaOH solution. Filter any insoluble material present, heavy metal oxides, through a coarse filter paper (e.g. Whatman number 41). Wash the precipitate with 0.6 M NaOH and collect filtrate and washings in a 50 mL volumetric flask. Dilute the sample to exactly 50 mL with distilled water.

The bottom of the *electrolysis cell* is covered with a layer of mercury. Contact between the mercury and the potentiostat is made through a platinum wire either sealed in the bottom of the electrolysis vessel or dipping into the mercury from above. A magnetic stir bar rests on the mercury surface. Exactly one mL of the sample solution is transferred to the electrolysis cell. Add 10 mL supporting electrolyte. (Prepare supporting electrolyte solution by dissolving 5.05 g oxalic acid, $H_2C_2O_4$, in 150 mL distilled water. Carefully add 15 mL 18 M H_2SO_4, cool, and adjust the pH to 2.1 by the slow addition of 15 M ammonia, NH_3.) If the pH of the solution is not 2.1, add dropwise 0.5 M H_2SO_4 until pH 2.1 is achieved. Deaerate the sample for approximately 7 min with a stream of pure nitrogen. Insert the working and reference electrodes, start the stirrer, and begin the electrolysis. Initial electrolysis current is generally between 15–50 mA. Continue the electrolysis until the current has dropped to approximately 20 μA. Electrolysis time is from 10–20 min.

If the potentiostat used in the experiment has a built-in integrator, the integrator can be calibrated (by substituting a standard resistor and constant potential source for the sample cell) to read directly a number proportional to milliequivalents molybdenum present in the sample. If manual plotting is employed (current decay versus time), it is necessary to determine the area under the current-time curve. This area represents the quantity of electricity (coulombs) needed in the reduction of molybdenum(VI) to molybdenum(V). Once the quantity of electricity is known, Faraday's relation (96,487 coulombs corresponds to one equivalent of reducible species) is used to calculate the amount of molybdenum present. Those unfamiliar with calculations pertaining to controlled potential coulometry should consult an appropriate reference [13]. Correct the amount of molybdenum present for dilutions made during sample preparation and report the percent Mo present in the original sample.

References

1. Manzone, M.G., Briggs, J.Z.: Molybdenum, Less Common Alloys of Molybdenum, 1962, New York, Climax Molybdenum Company
2. Jaffee, R.I., Promisel, N.E.: The Science, Technology and Application of Titanium, 1970, Oxford, Pergamon Press
3. Betteridge, W.: Nickel and Its Alloys, 1977, Plymouth, England, MacDonald and Evans
4. Yih, S.W.H., Wang, C.T.: Tungsten: Sources, Metallurgy, Properties, and Applications, 1979, New York, Plenum Press
5. Saltykova, A.M., Davidovich, N.K., Melamed, Sh.G.: Zh. Anal. Khim., 27(6), 1216 (1972)
6. Montgomery, E.L.: Anal. Chim. Acta, 121, 85 (1980)
7. Codell, M., Mikula, J.J., Norwitz, G.: Anal. Chem., 25, 1441 (1953)
8. Scarborough, J.M.: Anal. Chem., 41, 250 (1969)
9. Pashchenko, E.N., Mal'tsev, V.F., Zherikova, A.N.: Zh. Anal. Khim., 31(2), 400 (1976)
10. Elinson, S.V., Savvin, S.B., Nezhnova, T.I.: Zh. Anal. Khim., 22(4), 531 (1967)
11. Rigdon, L.P., Harrar, J.E.: Anal. Chem., 40, 1641 (1968)
12. Harrar, J.E., Behrin, E.: Anal. Chem., 39, 1230 (1967)
13. Rechnitz, G.A.: Controlled-Potential Analysis, 1963, New York, Macmillan

Chapter 17

Molybdenum Compounds

As a transition element in group VIB of the periodic table of chemical elements, molybdenum exhibits various oxidation states, plus six being the highest and most common [1]. Molybdenum in the $+5$, $+4$, $+3$, and $+2$ states is known although the lower states are progressively unstable and tend to revert to molybdenum(VI).

Molybdenum(VI) oxide is the starting material for many molybdenum compounds and, in addition, is itself useful in various applications. The properties of molybdenum(VI) oxide, MoO_3, are compiled as part of an overall source of information on metal oxides [2]. Commercially MoO_3, ammonium paramolybdate, $(NH_4)_6Mo_7O_{24} \cdot 4H_2O$, and sodium molybdate, $Na_2MoO_4 \cdot 2H_2O$, are the generally available forms of molybdenum. Each of these has been used as a standard for preparing calibration curves in analysis of molybdenum containing materials. The ammonium and sodium salts are generally not dried for this purpose rather, as reagent grade chemicals, used directly. Thermal decomposition curves for ammonium paramolybdate, indicating liberation of water and ammonia when heated, are available [3, 4]. For the most exacting work, however, these compounds should be standardized, preferably by a gravimetric procedure, to obtain an absolute value for their molybdenum content. Table 17-1 lists procedures for determining impurities in each of these compounds.

Besides the molybdenum compounds of analytical signficance discussed in earlier chapters, molybdenum compounds find applications in many commercial enterprises. Metal molybdates are used in paints as corrosion inhibitors and pigments. Soluble molybdates are found in circulating water systems as *corrosion inhibitors*. Molybdenum compounds function as catalysts in hydrodesulfurization of petroleum and in certain polymerization reactions. *Molybdenum containing enzymes* are necessary in utilization of nitrogen by plants and molybdenum is an essential element for animals and man. *Molybdenum disulfide*, MoS_2, is used as a lubricant. The analyses of commercial products, lubricants, catalysts, etc. for molybdenum are discussed in chapter 20.

One common and widely studied class of molybdenum compounds is the *heteropoly complexes* [17–19]. Of interest not only for their unique and unusual chemistry, they are of industrial importance as pigments and catalysts. They find application in analyses because of their ion exchange properties [20–22] and as analytical reagents [23, 24]. There are many types of heteropoly complexes, molybdenum being one of many elements capable of forming heteropoly species. All heteropoly complexes contain a central metal element bound to several second metal species which themselves are bound to oxygen. Perhaps the most widely studied heteropoly ion is *12-molybdophosphate ion*, $[PMo_{12}O_{40}]^{3-}$ containing a central

Table 17-1. Determination of Diverse Elements in Selected Molybdenum Compounds

Element	Procedure	Ref.
molybdenum oxide		
Al, Ag, As, Bi, Cd, Co, Cr, Cu, Fe, Mg, Ni, Pb, Sb, Si, Sn	emission spectroscopy	[5]
Al, B, Be, Bi, Ca, Cd, Co, Cr, Cu, Fe, Hf, Mg, Mn	emission spectroscopy	[6]
Ag, As, Bi, Cd, Cu, Fe, Pb, Sb, Sn, Te, Ti, W, Zn	emission spectroscopy	[7]
Cu, Zn	atomic absorption	[8]
K	flame emission	[9]
Li, Ti, Zr	emission spectroscopy	[10]
Si	colorimetry benzidine	[11]
ammonium paramolybdate		
As, Bi, Cu, Fe, Pb, Zn	gravimetry co-precipitate with $SnO_2 \cdot xH_2O$	[12]
$(NH_4)_2MoO_4$	laser Raman spectrometry	[13]
Si	colorimetry molybdosilicate	[14]
V	emission spectroscopy	[15]
sodium molybdate		
Fe	colorimetry bathophenanthroline	[16]
Si	colorimetry molybdosilicate	[14]

phosphorus ion surrounded by four equivalent tetrahedral molybdenum clusters. Series of heteropoly ions exist with ratios other than 1:12. Mixed heteropoly ions also occur in which not all of the surrounding metal ions are the same element. Heteropoly species are identified by their high molecular weight, water solubility, instability in basic solution, ion association complexes with positively charged cations, and their solubility in various organic solvents. Upon addition of a reducing agent, partial reduction of some of the metal ions occurs leading to a more intensely colored species. For 12-molybdophosphate reduction yields the characteristic heteropoly blue complex commonly used for colorimetric determination of phosphorus. Table 17-2 lists some metal ions and organic molecules determined by measurement of a heteropoly species.

Lorber and Müller [8] – MoO_3: Place 5 g molybdenum(VI) oxide sample in a teflon beaker. Cover the sample with 15 mL distilled water and add 3 g solid sodium hydroxide pellets, NaOH. Dissolve the material carefully as the sample may spatter because of the heat generated in dissolving NaOH. Rinse the sides of the beaker with water. Slowly add 5–6 mL 12 M hydrochloric acid, HCl, to neutralize the sample. pH of the final solution should be between 6–7. Depending upon the procedure to be followed, additional acid may be required. Dilute the sample to 100 mL in a volumetric flask.

Bourret, Lecuire, and Weis [103] describe a method for *determination of lower molybdenum oxides in the presence of molybdenum(VI) oxide*. They apply their procedure to determining the purity of molybdenum(IV) oxide, MoO_2, to the amount of MoO_2 present in a mixed MoO_2–MoO_3 sample, and to the purity of powdered molybdenum metal, $Mo°$. Lower molybdenum oxides react with periodate ion in neutral solution. After reaction and upon acidification, to pH 3 and in the presence

Table 17-2. Determination of Various Substances Through Heteropoly Formation with Molybdenum

Substance Determined	Heteropoly Species	Measurement	Reference
As	12-molybdoarsenate	gravimetry isoquinoline	[25]
As	12-molybdoarsenate	colorimetry reduced heteropoly	[26, 27, 28, 29]
B hydrides	12-molybdophosphate	colorimetry B_xH_y reduces heteropoly	[30]
Bi	18-molybdobismuthophosphate	colorimetry reduced heteropoly	[31]
Ce	12-molybdocerate	colorimetry molybdenum following precipitation	[32]
Cs	12-molybdophosphate	colorimetry reduced heteropoly following precipitation	[33]
Cs	12-molybdophosphate	potentiometry heteropoly ion electrode	[34]
Co	8-molybdo-4-tungstosilicate	colorimetry Co(II) reduces heteropoly	[35]
Cu	12-molybdophosphate	colorimetry $[Cu(CN)_2^-]$ reduces heteropoly	[36]
Fe	6-molybdo-6-tungstophosphate	colorimetry Fe(II) reduces heteropoly	[37]
Ga	11-tungsto-1-molybdogallate	colorimetry reduced heteropoly	[38]
Ge	12-molybdogermanate	colorimetry reduced heteropoly following precipitation crystal violet	[39]
Ge	12-molybdogermanate	colorimetry	[40]
Ge	12-molybdogermanate	colorimetry reduced heteropoly	[41]
Hf	12-molybdohaftnate	colorimetry reduced heteropoly	[42, 43]

Table 17-2 (continued)

Substance Determined	Heteropoly Species	Measurement	Reference
Nb	10-molybdo-2-niobophosphate	colorimetry	[44]
Nb	10-molybdo-2-nibiophosphate	colorimetry reduced heteropoly	[45, 46]
Ni	(not available)	colorimetry reduced heteropoly	[47]
P	12-molybdophosphate	gravimetry 8-hydroxyquinoline	[48, 49]
P	12-molybdophosphate	gravimetry quinoline	[50]
P	12-molybdophosphate	colorimetry reduced heteropoly	[51, 52, 53, 54, 55, 56]
P	11-molybdo-1-vanadophosphate	uv absorbance following extraction	[57, 58]
P	12-molybdophosphate	colorimetry follow extraction with complexing agent	[59, 60, 61]
P	11-molybdo-1-antimonophosphate	emission spectroscopy Mo following extraction	[62]
P	12-molybdophosphate	anodic stripping voltametry	[63]
Rb	10-molybdo-2-vanadophosphate	gravimetry Rb^+	[64]
Se	(not available)	colorimetry	[65]
Si	12-molybdosilicate	gravimetry pyridine	[66]
Si	12-molybdosilicate	gravimetry quinoline	[67]
Si	12-molybdosilicate	colorimetry	[68, 69]
Si	12-molybdosilicate	colorimetry reduced heteropoly	[70, 71, 72]
Si	12-molybdosilicate	colorimetry following extraction with complexing agent	[73, 74, 75, 76]
Ta	(not available)	emission spectroscopy Ta	[77]
Te	11-molybdo-1-tellurophosphate	colorimetry	[78]
Th	(not available)	colorimetry reduced heteropoly	[79]

Table 17-2 (continued)

Substance Determined	Heteropoly Species	Measurement	Reference
Ti	(not available)	colorimetry	[80]
Ti	(not available)	uv absorbance	[81]
Ti	(not available)	colorimetry reduced heteropoly	[82]
Tl	12-molybdophosphate	uv absorbance following Tl^+ precipitation	[83]
V	11-molybdo-1-vanadophosphate	uv absorbance	[84]
W	(not available)	gravimetry β-naphthaquinoline	[85]
W	(not available)	colorimetry reduced heteropoly	[86]
W	(not available)	gravimetry reduced heteropoly	[87]
Y	(not available)	colorimetry reduced heteropoly	[88]
Zr	12-molybdo-1-zirconophosphate	colorimetry	[89]
aldehydes, ketones	12-molybdosilicate	visual color	[90]
alkaloids	12-molybdosilicate	colorimetry	[91]
aminocarbox-ylic acids	(not available)	visual color	[92]
aromatic amines	12-molybdophosphate	colorimetry	[93]
catechol	12-molybdophosphate	colorimetry	[94]
cholesterol	12-molybdo-6-tungstophosphate	colorimetry reduced heteropoly	[95]
egg albumin	12-molybdophosphate	potentiometry	[96]
hydrazine	6-molybdo-6-tungstophosphate	colorimetry N_2H_4 reduces heteropoly	[97]
nonionic surfactants	12-molybdophosphate	potentiometry	[98]
penicillin	12-molybdophosphate	colorimetry	[99]
reducing sugars	12-molybdophosphate	colorimetry sugar reduces heteropoly	[100]
tissue staining	12-molybdophosphate	visual color	[101, 102]

of added excess molybdate ion (to complex unreacted IO_4^-), the iodate formed selectively reacts with added KI solution to produce iodine. Periodate present, when complexed with molybdate, does not react with iodide ion. Liberated iodine is titrated with standard thiosulfate solution. If, upon a separate aliquot, molybdenum concentration is also found, then the oxidation number of the particular lower oxide present can be calculated. They achieved purity values of 100.2 ± 0.7 (standard deviation) per cent for the mean of three measurements upon a sample of MoO_2 and 99.4 ± 0.6 standard deviation) per cent for the mean of two measurements upon a sample of powdered molybdenum metal. Other reducing agents present in the sample that react with periodate interfere with this procedure.

Place 40–50 mg sample, accurately weighed, into a 100 mL Erlenmeyer flask. Add 10.0 mL 5.0% (w/v) sodium periodate solution, $NaIO_4$. Add 5.0 mL 1% (w/v) sodium bicarbonate solution, $KHCO_3$, and 25 mL distilled water.

$$2\,MO_{3-x/2} + \times IO_4^- + 2\,H_2O \rightarrow 2\,MoO_4^{2-} + \times IO_3^- + 4\,H^+$$

pH control is critical for should the acidity increase (because of generation of H^+ ions during the course of the reaction), as indicated by a coloration of the solution, additional $KHCO_3$ solution is required. Maintain the sample at approximately 60 °C with constant stirring until the sample dissolves. In the case of MoO_3 samples, where the matrix is already in its highest oxidation state, it will be necessary to filter the sample after all the lower state oxides are believed in solution. Use a fine filter paper (e.g. Whatman number 42). Thoroughly wash the residue with distilled water and collect filtrate and washings. Once dissolved, or filtered, transfer the sample to a 100 mL volumetric flask and dilute with distilled water to exactly 100 mL.

To determine the amount of iodate formed, transfer exactly 40.0 mL sample to a 250 mL Erlenmeyer flask. Add 30 mL 40% (w/v) sodium molybdate, $Na_2MoO_4 \cdot 2H_2O$. Add slowly and with stirring pure chloroacetic acid, $CH_2ClCOOH$, until the pH of the solution is 3.0. Use a pH meter in making this measurement. Add, with stirring, 2 g solid potassium iodide, KI. Stir to dissolve KI. Iodine formed from the reaction of iodate and iodide in acid solution is titrated with 0.0500 M sodium thiosulfate, $Na_2S_2O_3$. ($Na_2S_2O_3$ solution is itself standardized by reacting a weighed amount of potassium iodate, KIO_3, with excess KI under the conditions described for this experiment.) Near the end point of the titration, iodine color nearly gone, add 2–3 drops 1% (w/v) starch solution. (Prepare by forming a paste with 1 g starch and 5–10 mL distilled water. Transfer the paste to a beaker containing boiling water. Stir the mixture and dilute to 100 mL with additional water. Use the clear supernatant liquid.) The end point is reached upon disappearance of the blue iodine-starch color.

$$IO_3^- + 5\,I^- + 6\,H^+ \rightarrow 3\,I_2 + 3\,H_2O$$
$$I_2 + 2\,S_2O_3^{2-} \rightarrow S_4O_6^{2-} + 2\,I^-$$

To determine the amount of molybdenum oxidized by periodate, transfer exactly 20.0 mL sample to a 1,000 mL beaker. Add 180 mL distilled water. Slowly add 1 M perchloric acid, $HClO_4$, until the solution is pH 1. Use a pH meter. Add, with stirring, 0.5 g potassium iodide, KI, solid. The KI reduces IO_4^- present forming I_2. Add 0.2 M $Na_2S_2O_3$ to destroy I_2. Add 5 mL additional 1 M $HClO_4$ and, if traces of I_2 still persist, additional 0.2 M $Na_2S_2O_3$. Dilute the sample to 500 mL as indicated by the mark in the side of the beaker. Add solid sodium acetate, $NaC_2H_3O_2$. Continue adding $NaC_2H_3O_2$ until the pH of the solution is 5.3 as indicated by a pH meter. Heat the sample to 80 °C, add 2–3 drops 0.1% (w/v) xylenol orange indicator (dissolved in water), and slowly add 0.0200 M lead nitrate solution, $Pb(NO_3)_2$. (Standardize $Pb(NO_3)_2$ by titrating it against a weighed sample of sodium ethylenediaminetetraacetate, $Na_2H_2EDTA \cdot 2H_2O$, previously dried for 1 h at 80 °C. Conditions for standardization are the same as outlined here.) Lead molybdate, $PbMoO_4$, precipitates. Continue adding $Pb(NO_3)_2$ solution, with stirring, until the sample becomes violet, indicating an excess of lead ion being present. Add an additional 10.0 mL $Pb(NO_3)_2$ solution. Remove the sample from the hot plate. Check that the pH is still 5.3. If it is not, add additional solid $NaC_2H_3O_2$, returning the pH to 5.3. Allow the sample to set for two hours at ambient temperature. At the end of this time withdraw 25 mL of supernatant liquid containing excess $Pb(NO_3)_2$. Titrate this sample at room temperature with 0.0200 M EDTA solution. (Prepare by weighing the appropriate amount of $Na_2H_2EDTA \cdot 2H_2O$ dried at 80 °C for one hour and diluting to exactly 1 L in a volumetric flask.) The end poing occurs when the sample changes from violet to yellow.

First calculate the amount of molybdenum present in the known oxidation state. From the reaction with $Pb(NO_3)_2$ solution:

$$\text{mmol Pb added} = \text{mL Pb added} \times \frac{0.0200 \text{ mmol Pb}}{\text{mL}} = A$$

$$\text{mmol Pb excess} = \text{mL EDTA} \times \frac{0.200 \text{ mmol EDTA}}{\text{mL EDTA}} \times \frac{1 \text{ mmol Pb}}{1 \text{ mmol EDTA}} \times \frac{500 \text{ ml}}{25.0 \text{ mL}} = B$$

$$\text{mmol Mo present} = A - B \text{ mmol Pb} \times \frac{1 \text{ mmol Mo}}{1 \text{ mmol Pb}} \times \frac{100 \text{ mL}}{20 \text{ mL}} = C$$

The amount of IO_3^- formed is also a measure of the amount of molybdenum in the lower oxidation state.

$$\text{mmol Mo} = C = \text{mL S}_2O_3^{2-} \times \frac{0.0500 \text{ mmol S}_2O_3^{2-}}{\text{mL S}_2O_3^{2-}}$$

$$\times \frac{1 \text{ mmol I}_2}{2 \text{ mmol S}_2O_3^{2-}} \times \frac{100 \text{ mL}}{40.0 \text{ mL}} \times \frac{1 \text{ mmol IO}_3^-}{3 \text{ mmol I}_2}$$

$$\times \frac{2 \text{ mmol MoO}_{3-x/2}}{x \text{ mmol IO}_3^-}$$

x represents the amount of iodate produced and is used in finding the number of oxygens associated with the lower oxidation state of molybdenum. Solve the above equations for x.

Once x is known, the particular oxide impurity present in a sample is known. To calculate per cent, purity:

$$\% \text{ purity} = \frac{\left(\begin{array}{c}\text{sample}\\\text{weight, mg}\end{array}\right) - \left(\begin{array}{c}\text{mmol}\\\text{lower oxide}\end{array}\right) \times \left(\begin{array}{c}\text{mol. wt.}\\\text{lower oxide, mg/mmol}\end{array}\right) \times 100}{\text{sample weight, mg}}$$

References

1. Stiefel, E.I.: Prog. Inorg. Chem., 22, 3 (1977)
2. Samsonov, G.V.: The Oxide Handbook 1982 New York, IFI/Plenum
3. Kiss, A.: Magy. Kem. Foly., 75, 302 (1969), Chem. Abstr., 71, 108722s (1969)
4. Park, I.-H.: Bull. Chem. Soc. Jpn., 45, 2739 (1972)
5. Molenda, K.: Pr. Inst. Mech. Precyz., 15(3), 47 (1967); Chem. Abstr., 69, 24181k (1968)

6. Morris, W.F., Worden, E.F.: Appl. Spectrosc., *25*, 305 (1971)
7. Delijska, A., Prǎvčeva, Kh.: Fresenius' Z. Anal. Chem., *281*, 281 (1976)
8. Lorber, K.E., Müller, K.: Mikrochim. Acta, II, 5 (1977)
9. Hegedus, A.J., Fukker, K.: Mikrochim. Acta, 357 (1962)
10. Razmadze, G.B., Khlystova, A.D., Artemova, T.N.: Vestn. Mosk. Univ., Khim., *17*(5), 581 (1976); Chem. Abstr., *86*, 132976h (1977)
11. Fukker, K., Hegedus, A.J.: Mikrochim. Acta, 226 (1961)
12. Papageorgios, P., Walczyk, W., Plonka, M.: Chem. Stosow., Ser. A, *12*(4), 469 (1968); Chem. Abstr., *70*, 92708h (1969)
13. Murata, K., Ikeda, S.: Bunseki Kagaku, *29*, 80 (1980)
14. Radovskaya, T.L., Barkovskii, V.F.: Zavod. Lab., *34*, 411 (1968)
15. Tsyganok, L.P., Chuiko, V.T., Reznik, B.E., Mazan, L.K., Stets', T.V.: Zavod. Lab., *39*, 169 (1973)
16. Galliford, D.J.B., Newman, E.J.: Analyst (London), *87*, 68 (1962)
17. Weakley, T.J.R.: Struct. Bonding (Berlin), *18*, 131 (1974)
18. Tsigdinos, G.A.: Topics Current Chem., *76*, 1 (1978)
19. Backer, P.A., Natarajan, L.V.: J. Chem. Educ., *56*, 642 (1979)
20. Buchtela, K., Lesegang, M.: Mikrochim. Acta, 67 (1965)
21. Pekárek, V., Veselý, V.: Talanta, *19*, 1245 (1972)
22. Murthy, T.S., Balasubramanian, K.R., Ananthakrishnan, M., Ramani, K.S., Varma, R.N., Anthony, M.C.: Indian AEC Bhabha At. Res. Cent., Rep. BARC-893, (1977); Anal. Abstr., *36*, 2B23 (1979)
23. Jean, M.: Ann. Chim. (Paris), [12]*3*, 470 (1948)
24. Halasz, A., Pungor, E.: Talanta, *18*, 557 (1971)
25. Lóránt, B.: Fresenius' Z. Anal. Chem., *274*, 125 (1975)
26. Duval, L.: Chim. Anal. (Paris), *51*, 415 (1969)
27. Malyutina, T.M., Savvin, S.B., Orlova, V.A., Mineeva, V.A., Kirillova, T.I.: Zh. Anal. Khim., *29*(5), 925 (1974)
28. Nakamura, E., Namiki, H.: Kogyo Yosui, *226*, 30 (1977); Chem. Abstr., *88*, 98648g (1978)
29. Ivanov, N.A.: Dokl. Bolg. Akad. Nauk, *32*, 1255 (1979); Chem. Abstr., *92*, 208432z (1980)
30. Hill, W.H., Merrill, J.M., Larsen, R.H.: Am. Ind. Hyg. Assoc. J., *20*, 5 (1959)
31. Hargis, L.G.: Anal. Chem., *41*, 597 (1969)
32. Shakhova, Z.F., Gavrilova, S.A.: Zh. Neorg. Khim., *3*, 1370 (1958)
33. Huey, F., Hargis, L.G.: Anal. Chem., *39*, 125 (1967)
34. Coetzee, C.J., Basson, A.J.: Anal. Chim. Acta, *57*, 478 (1971)
35. Kokorin, A.I., Radenko, S.K.: Uch. Zap. Kishinev. Gos. Univ. No.68, 45 (1963); Chem. Abstr., *63*, 2375c (1965)
36. Tobia, S.K., Gawargious, Y.A., El-Shahat, M.F.: Anal. Chim. Acta, *39*, 392 (1967)
37. Suteu, A., Maniu, C.: Rev. Roum. Chim., *13*, 1201 (1968); Chem. Abstr., *70*, 73896b (1969)
38. Tkach, V.I., Tsyganok, L.P.: Zh. Anal. Khim., *34*(10), 1943 (1979)
39. Mirzoyan, F.V., Tarayan, V.M., Hairiyan, E.Kh., Grigorian, N.A.: Talanta, *27*, 1055 (1980)
40. Jakubiec, R.J., Boltz, D.F.: Anal. Chem., *41*, 78 (1969)
41. Rudenko, V.K., Berezhnaya, Yu.F.: Zh. Neorg. Khim., *25*(9), 2697 (1980)
42. Shakhova, Z.F., Semenovskaya, E.N., Sokovikova, N.K.: Zh. Anal. Khim., *23*(8), 1164 (1968)
43. Clowers, C.C., Guyon, J.C.: Mikrochim. Acta, 989 (1969)
44. Shkaravskii, Yu.F.: Zh. Anal. Khim., *18*, 196 (1963)
45. Zaboeva, M.I., Barkovskii, V.F.: Zh. Anal. Khim., *17*, 955 (1962)

46. Hsi, K.-C., Chang, N.-Y., Tang, C.-T.: Fen Hsi Hua Hsueh, *8*, 400 (1980); Anal. Abstr., *42*, 5B139 (1982)
47. Heller, R.L., Guyon, J.C.: Talanta, *17*, 865 (1970)
48. Brabson, J.A., Edwards, O.W.: Anal. Chem., *28*, 1485 (1956)
49. Schulze, G., Noack, S.: Fresenius' Z. Anal. Chem., *294*, 117 (1979)
50. Schulze, G., Noack, S.: Fresenius' Z. Anal. Chem., *306*, 196 (1981)
51. Crouch, S.R., Malmstadt, H.V.: Anal. Chem., *39*, 1084 (1967)
52. Ohashi, K., Yasu, K., Suzuki, C., Yamamoto, K.: Bull. Chem. Soc. Jpn., *50*, 3202 (1977)
53. Ptushkina, M.N., Lebedeva, L.I., Petrokanskaya, I.Yu.: Zh. Anal. Khim., *35*(11), 2132 (1980)
54. Ohashi, K., Nakazawa, H., Enomoto, T., Yamamoto, K.: Bunseki Kagaku, *30*, 727 (1981)
55. Lueck, C.H., Boltz, D.F.: Anal. Chem., *28*, 1168 (1956)
56. Won, C.H.: Nippon Kagaku Zasshi, *85*, 908 (1964)
57. Jakubiec, R.J., Boltz, D.F.: Mikrochim. Acta, 181 (1969)
58. Kitagawa, H., Shibata, N.: Bunseki Kagaku, *8*, 302 (1959)
59. Ivanov, N.A.: Zh. Anal. Khim., *32*(9), 1688 (1977)
60. Krohn, J.: Fresenius' Z. Anal. Chem., *301*, 431 (1980)
61. Motomizu, S., Wakimoto, T., Tôei, K.: Anal. Chim. Acta, *138*, 329 (1982)
62. Miyazaki, A., Kimura, A., Umezaki, Y.: Anal. Chim. Acta, *138*, 121 (1982)
63. Fogg, A.G., Bsebsu, N.K.: Analyst (London), *106*, 369 (1981)
64. Krivy, I., Krtil, J.: Collect. Czech. Chem. Commun., *29*, 587 (1964)
65. Shakhova, Z.F., Gavrilova, S.A., Zakharova, V.F.: Vestn. Mosk. Univ., Ser. II, *21*, 64 (1966); Chem. Abstr., *66*, 82118v (1967)
66. Harzdorf, C.: Fresenius' Z. Anal. Chem., *227*, 96 (1967)
67. MacDonald, A.M.G., Van der Voort, F.H.: Analyst (London), *93*, 65 (1968)
68. Truesdale, V.W., Smith, C.J.: Analyst (London), *100*, 203 (1975)
69. Beshikdash'yan, M.T., Vasil'eva, M.G.: Zh. Anal. Khim., *36*(6), 1082 (1981)
70. Andersson, L.H.: Ark. Kem., *19*, 235 (1962)
71. Kakita, Y., Goto, H.: Talanta, *14*, 543 (1967)
72. Tarasova, N.S., Dorokhova, E.N., Alimarin, I.P.: Zh. Anal. Khim., *36*(3), 459 (1981)
73. Golkowska, A., Pszonicki, L.: Talanta, *20*, 749 (1973)
74. Mirzoyan, F.V., Tarayan, V.M.: Zh. Anal. Khim., *35*(7), 1293 (1980)
75. Mirzoyan, F.V., Tarayan, V.M., Karapetyan, Z.A.: Arm. Khim. Zh., *34*(2), 122 (1981); Anal. Abstr., *41*, 3B120 (1981)
76. Xi, G., Zhang, N.: Fen Hsi Hua Hsueh, *9*, 6 (1981); Anal. Abstr., *41*, 3B121 (1981)
77. Semenenko, K.A., Tarasevich, N.I.: Zh. Anal. Khim., *18*, 88 (1963)
78. K'u, T.L., Sudakov, F.P., Shakhova, Z.F.: Zh. Anal. Khim., *19*(8), 968 (1964)
79. Madison, B.L., Guyon, J.C.: Anal. Chem., *39*, 1706 (1967)
80. Shakhova, Z.F., Semenovskaya, E.N.: Vestn. Mosk. Univ., Ser. II, *23*, 122 (1968); Chem. Abstr., *69*, 92727t (1968)
81. Shkaravskii, Yu.F.: Zh. Anal. Khim., *19*(3), 320 (1964)
82. Veĭtsman, R.M.: Zavod. Lab., *25*, 408 (1959)
83. Hargis, L.G., Boltz, D.F.: Anal. Chem., *37*, 240 (1965)
84. Jakubiec, R.J., Boltz, D.F.: Anal. Chim. Acta, *43*, 137 (1968)
85. Latichevskii, I.K., Polotebnova, N.A.: Khim. Fiz.-Khim. Metody Issled. Soedin. 1980 Kishinev USSR, Izd. Shtiintsa; Chem. Abstr., *94*, 95266u (1981)
86. Guyon, J.C., Marks, J.Y.: Anal. Chem., *40*, 837 (1968)
87. Reznik, B.E., Ganzburg, G.M., Mal'tseva, G.V.: Zh. Anal. Khim., *23*, 1848 (1968)
88. Madison, B.L., Guyon, J.C.: Anal. Chim. Acta, *42*, 415 (1968)
89. Murata, K., Yokoyama, Y., Ikeda, S.: Anal. Chim. Acta, *48*, 349 (1969)

90. Billman, J.H., Borders, D.B., Buehler, J.A., Seiling, A.W.: Anal. Chem., *37*, 264 (1965)
91. Kramarenko, V.P., Rohach, Z.S.: Farm. Zh. (Kiev), *16*, 26 (1961); Chem. Abstr., *56*, 11706i (1962)
92. L'Annunziata, M.F., Fuller, W.H.: J. Chromatogr., *34*, 270 (1968)
93. Coeur, A., Boucherle, A.: Bull. Trav. Soc. Pharm. Lyon, *7*, 115 (1963); Chem. Abstr., *60*, 13886d (1964)
94. Strachota, J., Kotásek, Z.: Chem. Listy, *52*, 1093 (1958); Chem. Abstr., *52*, 15348b (1958)
95. Negrin, A.: Clin. Chem., *15*, 829 (1969)
96. Bobtelsky, M., Barzily, I.: Anal. Chim. Acta, *35*, 520 (1966)
97. Suteu, A., Maniu, C.: Rev. Roum. Chim., *15*, 119 (1970); Chem. Abstr., *73*, 21100e (1970)
98. Vinnikov, Yu.A., Kostareva, L.A., Gruzdova, N.S.: Zh. Anal. Khim., *36*(3), 465 (1981)
99. Deshpandu, G.R., Karmarkar, S.S.: Hind. Antibio. Bull., *3*, 174 (1961); Chem. Abstr., *55*, 25159d (1961)
100. Lo, C.-P., Chu, L.J.-Y.: Ind. Eng. Chem., Anal. Ed., *16*, 637 (1944)
101. Vignoli, L., Cristau, B., Pfister, A.: Bull. Soc. Pharm. Marseilles, *4*(13), 71 (1955); Chem. Abstr., *50*, 531g (1956)
102. Fritz, H.: Mikroskopie, *19*, 110 (1964)
103. Bourret, P., Lecuire, J.M., Weis, C.: Chim. Anal. (Paris), *52*, 48 (1970)

Chapter 18

Animal, Plant, and Soil Samples

Molybdenum is an essential element in biological systems, both animal and plant. Its role in life processes is only now starting to be understood and the need for determination of molybdenum in biological materials grows as further study of its significance proceeds.

1. Animal Samples

Molybdenum, to varying degrees, is found in most animal matter [1]. Highest concentrations are frequently present in the liver, but it is also present in other parts of the body [2]. Cited as part of an extensive listing of trace elements in human blood, molybdenum values ranged from 0.58 to 257 ng/mL [3]. The authors point out, however, the possibility of contaminated samples and they favor those studies reporting molybdenum values near the lower end of this range. More recent studies, too, support a low ng Mo/mL value for blood molybdenum [4, 5].

Molybdenum is associated with the enzyme *xanthine oxidase* which, in humans, is necessary for proper formation of uric acid. Other enzymes with which molybdenum is found are xanthine dehydrogenase, aldehyde oxidase, and sulfite oxidase [6, 7]. An excessive amount of molybdenum in animals is also accompanied by copper deficiency [7]. Although considered less toxic than chromium, *excessive amounts of molybdenum are hazardous* to animals in several ways. Molybdenum compounds, molybdenum oxide, chloride, and ammonium molybdate are more hazardous than molybdenum itself. Possible symptoms of excessive molybdenum include gasterointestinal disorders, diarrhea, emaciation, and possibly death [8, 9]. Elevated molybdenum blood levels have been linked to the presence of certain cancers in man [10]. Exposure to molybdenum containing dust, for example, in mining and purification of molybdenum ores, can lead to breathing disorders, *molybdenosis*. Present threshold limits for molybdenum in air are 5 mg/m^3 soluble molybdenum and 15 mg Mo/m^3 insoluble molybdenum [11].

Animal samples are readied for molybdenum determination by either wet acid digestion or dry ashing to decompose organic matter. Choice of acid or acid mixture and drying conditions vary widely. One must be careful to avoid loss of trace molybdenum through volatilization of molybdenum compounds during the decomposition process. Suggested temperatures for dry ashing and conditions for wet digestion with H_2SO_4–$HClO_4$–HNO_3 (1:1:3 v/v) mixtures are referenced in the abstract literature for trace elements in blood [12]. When only limited amounts of sample are available, burning in a stream of oxygen with collection in liquid air is suggested as an alternate sample treatment [13].

Norheim and Waasjø [14] – *liver*: Place 5 g sample in a 250 mL beaker. Add 5 mL 18 M sulfuric acid, H_2SO_4, and 10 mL 15 M nitric acid, HNO_3. Heat the sample on a hot plate stirring frequently to assure contact between sample and acids. Add an additional 5 mL 15 M HNO_3 as charring begins. After reaction ceases add 1 mL 70% (w/v) perchloric acid, $HClO_4$. Continue heating until the sample is clear. Increase the temperature, evaporating the solution, until fumes of sulfur trioxide, SO_3, are observed. Cool the sample, add 10 mL distilled water, and again evaporate the sample to fumes of SO_3. Transfer the sample to a 100 mL volumetric flask. Rinse the beaker thoroughly with distilled water and add the rinsings to the flask. Dilute the sample with water, or appropriate acid solution, until the final volume is exactly 100 mL.

Bentley, Markowitz, and Meglen [5] – *serum or urine*: With a micropipet place a measured amount of sample, between 0.30 and 1.00 mL, in a borosilicate vial. Add 3 drops 15 M nitric acid, HNO_3. Mix the contents of the vial and place it in an oven at 70 °C for several hours or overnight. The dried sample is placed in a commercial low temperature ashing device (e.g. Tracerlab Model 600) for 8–24 h. The ash should be nearly white in color. Add 1 drop 15 M HNO_3 to wet the ash and repeat the drying procedure, first at 70 °C for several hours and then for 8–24 h in the low temperature asher. The residue is now dissolved and made up to its original volume (0.30 to 1.00 mL) by careful addition of 6 M hydrochloric acid, HCl. Be sure to dissolve all the ash particles, swirling the vial if necessary to achieve contact between acid and particles adhering to the sides of the container. The authors intend the sample be used for molybdenum determination by flameless atomic absorption spectrometry.

Bentley, Markowitz, and Meglen [5] – *red blood cells*: The end of a disposable pipet tip is cut off to provide an opening of about 2 mm and weighed. An amount of sample, about 0.5 mL, is placed in the pipet tip and weighed to ascertain the sample weight. The aliquot, or a similar one, is placed in a 2 mL borosilicate vial. Rinse the pipet tip thoroughly with three 0.5 mL portions of distilled water and collect all washings with the sample in the vial. Add 3 drops 15 M nitric acid, HNO_3. Allow the sample to set at room temperature for 12–16 h. The sample should, at the end of this time, be a gray-brown color. Transfer the vial to a 70 °C oven for an additional 12 h, or until dry. The dry sample is placed in a low temperature ashing device (e.g. Tracerlab Model 600) and ashed for 16 h. Repeat the cycle, 1 drop 15 M HNO_3, 70 °C drying for 12 h, low temperature ashing, until the ash appears white. Slowly and carefully add about 0.5 mL 15 M HNO_3 to dissolve the sample. Swirl the vial to assure minute particles of ash on the sides of the vial are dissolved. The authors intend the sample be used for molybdenum determination by flameless atomic absorption spectrometry.

Christian and Patriarche [4] describe a procedure for determination of nanogram Mo/mL in *blood or urine* samples based upon the catalyzing effect of minute amounts of molybdenum upon the reduction of selenium(VI), SeO_4^{2-}, to elemental selenium, Se°. The red colloidal Se° suspension is monitored by reading absorbance values from a colorimeter set at 390 nm. Because of the catalytic action of molybdenum, the time between start of the reaction and measurement must be constant for all samples. The alternative, but more involved procedure, would be to monitor each standard and sample over a period of time, plot the change, and from the slope of each plot prepare a calibration curve. Linearity of the authors' calibration curve is observed for up to 0.5 µg Mo/25 mL sample. Following dry ashing, samples are extracted to separate molybdenum from major interfering substances prior to the catalytic reaction. As, V, Fe, Te, Ge, and Co are potential interferences in this procedure. Tellurium and some arsenic are lost during the drying process. The other elements, to the extent they appear in the samples studied, seemed to present no problems. Reagent purity is a concern with determination at the nanogram level and preparation of a new calibration curve for each new batch of reagents is suggested. Triply distilled and deionized water was used throughout the experiment. The authors report on replicate samples a molybdenum content (and standard deviation) of 7 ± 5 ng Mo/mL (human plasma), and 5 ± 2 ng Mo/mL (human blood). For human urine they found 280 and 320 ng Mo excretion for a twenty-four hour period.

A *molybdenum standard solution* is prepared by dissolving 2.52 g sodium molybdate, $Na_2MoO_4 \cdot 2H_2O$, in water and diluting to a final volume of 1 L in a volumetric flask. Exactly 1.00 mL of this solution is transferred to a second 1 L volumetric flask. Dilute this sample to one liter with distilled water and thoroughly mix the contents of the flask. The diluted stock solution contains 1.00 µg Mo/mL.

Transfer to separate 30 mL separatory funnels 0.10, 0.20, 0.30, and 0.40 mL diluted stock solution. Add to each funnel 8.0 mL 6 M hydrochloric acid, HCl. To each of two other 30 mL separatory funnels, add 8.0 mL 6 M HCl. A double blank is recommended because of the minute concentration levels being measured and the chance for contamination from reagents and glassware.

For whole blood, red cells, plasma, or serum, place 10.0 mL sample in a porcelain crucible. For urine samples, place 1.00 mL sample in a porcelain crucible. Carefully evaporate the sample over a low flame. Continue heating until smoke clears and only charred sample remains. Place the charred sample in a furnace at 550 °C for three hours. Remove the sample from the furnace. When cool, add 4.0 mL 6 M HCl. Transfer the dissolved sample to a 30 mL separatory funnel. Rinse the crucible with an additional 4.0 mL 6 M HCl and add the rinsing to the separatory funnel.

To each separatory funnel, containing either blank, standard, or sample, add 10 mL n-pentyl acetate, $CH_3COOC_5H_{11}$. Shake each funnel for 10 min to extract molybdenum into the organic phase. After the layers separate, discard the aqueous layer. Add, to each funnel, 10 mL water. Shake each funnel again for several minutes to transfer molybdenum back to the aqueous phase. After layer separation transfer each aqueous sample to a separate, labeled, 25 mL volumetric flask. Add to each flask 4.00 mL 6 M HCl and 2.00 mL purified 0.45 M sodium selenate solution. (To avoid excessively high blank values, it is necessary to purify the Na_2SeO_4 solution. Dissolve 8.50 g solid Na_2SeO_4 in water and dilute to 100 mL final volume with additional water. Add 5 mL 0.1 M potassium permanganate solution, $KMnO_4$. Heat the sample to boiling and add 1 mL 0.1 M iron(II) ammonium sulfate, $FeSO_4 \cdot (NH_4)_2SO_4$. Continue boiling the sample for 10 min. Cool and filter the solid manganese(IV) oxide, MnO_2, formed through a course filter paper (e.g. Whatman number 41). Do not wash the residue containing possible impurities. Collect the selenate filtrate for use in this experiment. Coloration of the solution by excess permanganate is not detrimental to the determination of molybdenum.) Add 0.5 mL 0.25% (w/v) gum arabic solution (dissolve in water). Each blank, standard, and sample is individually timed from the moment tin(II) chloride solution is added. Add with pipet 4.00 mL 0.4 M $SnCl_2$ solution. (Prepare by dissolving 9.0 g $SnCl_2 \cdot 2H_2O$ in 6 M HCl. Dilute to 100 mL with additional 6 M HCl.) Immediately start a timer. Add sufficient distilled water to make 25 mL total volume, mix the contents of the flask, and place it in a constant temperature bath at 25 ± 0.5 °C. After exactly 10.0 min, remove the flask, transfer an aliquot to a colorimeter cell and read the absorbance at 390 nm. If a double beam colorimeter is used, place water in the reference cell. Zero single beam colorimeters for 100% T with water. Repeat this procedure for each blank, standard, and sample.

Correct, by substraction, each absorbance value for the standards and sample by the mean absorbance value of the two blank samples. Prepare a calibration curve of absorbance versus molybdenum content for each standard (0.10 through 0.40 µg Mo/25 mL). From the corrected absorbance of the unknown read its molybdenum content from the calibration curve. If necessary correct this reading for sample volume to obtain µg Mo/mL sample.

2. Plant Samples

Molybdenum is a necessary component of the enzyme *nitrogenase* responsible for conversion of elemental nitrogen to ammonia by certain soil bacteria [15–20]. Without this process available nitrogen for plant growth would be sharply reduced. The enzyme consists of two components, one containing both iron and molybdenum (32 atoms Fe, 2 atoms Mo, molecular weight 230,000) and the other iron only (4 atoms Fe, molecular weight 65,000). The detailed mechanism by which this conversion takes place and the actual role of molybdenum are presently a very active area of investigation. The overall importance of molybdenum to agriculture has been assessed [21, 22] and earlier procedures for determination of molybdenum in plants reviewed [23, 24]. Present in the range µg Mo/g dried plant material, or less, sensitive methods are necessary to accurately determine molybdenum content of plants and careful sample preparation is needed to avoid molybdenum loss. Preconcentration and/or separation from accompanying interferences by extraction or other means are also frequently employed. Both wet and dry ashing of plant samples have preceeded actual molybdenum determination [25].

Benne and Linden [26]: Place the air dried sample in an oven at 105 °C for several hours, or overnight, to remove moisture. Grind the sample to a fine powder in a mortar. Weigh between 1 and 5 g sample and place it in a 200 mL tall form beaker. For 1, 2, or 5 g samples add, respectifely, 10, 15, or 35 mL 15 M nitric acid, HNO_3. Cover the beaker with a watch glass and after initial reaction subsides, about 15 min, heat the sample on a hot plate to near 100 °C. Maintain this condition for 2 h or until most of the plant ash is dissolved. Remove the beaker from the hot plate if frothing occurs. Cool the sample to room temperature, add 2–3 mL 15 M HNO_3, and 6 mL 70% (w/v) perchloric acid, $HClO_4$. Cover the beaker, return it to the hot plate, and boil the solution until dissolution is complete. A colorless or pale yellow solution should be present. It may be necessary to add additional HNO_3 and/or $HClO_4$ during the boiling. After the sample is dissolved, support the cover glass above the rim of the beaker with glass hooks and apply heat to evaporate the sample to near dryness. Be careful when evaporating perchloric acid solutions as explosions are possible if organic matter is allowed in contact with hot, concentrated $HClO_4$ solutions. Cool the beaker. Rinse the cover glass and sides of the beaker with a few mL of distilled water, allowing the washings to collect in the beaker. Heat the sample briefly to boiling, cool, and again rinse the container with a small amount of distilled water. Add 6 M hydrochloric acid, HCl, or other appropriate acid, depending upon subsequent sample treatment. Quantitatively transfer the sample to a 100 mL volumetric flask and dilute it to exactly 100 mL with additional acid.

Heanes [27]: Three mL acidified potassium hydrogen sulfate solution (Prepare by diluting 25 mL 15 M nitric acid, HNO_3, and 12.5 g $KHSO_4$ to a final volume of 100 mL with distilled water.) is placed in a borosilicate bottle fitted with an acid resistant screw cap. A volume mark at 55 mL should be scratched upon the side of the bottle. Add a weighed 4 g sample of ground plant material previously dried at 105 °C for several hours to remove moisture. Add 3 mL additional acidified $KHSO_4$ solution. Place the uncapped bottle in a muffle furnace capable of temperature programming and preheated to 120 °C. Increase the furnace temperature from 120 °C to 175 °C over a period of one hour. If the sample solution has not yet evaporated, maintain the 175 °C temperature until it does. The furnace door is ajar to allow for escape of moisture. Program a temperature increase from 175 °C to 350 °C over a one hour period. If the sample is not completely charred, maintain 350 °C until all additional charring ceases. Program a temperature increase from 350 °C to 550 °C over a 1.5 h period. Sulfuric acid and residual organic compounds are now removed. Maintain a temperature of 550 °C with the furnace door closed and with a stream of pure oxygen, O_2, flowing through the muffle, at 100 mL/min, for ten hours. Allow the furnace to cool below 300 °C before opening the door. Turn off the oxygen flow and remove the sample when the temperature drops below 150 °C. When the sample is at room temperature, add 3 mL 25% (v/v) nitric acid – 25% (v/v) sulfuric acid mixture. (Prepare by adding first 25 mL 15 M HNO_3 to 50 mL distilled water. Then add, carefully, 25 mL 18 M H_2SO_4 followed by sufficient distilled water, after the sample has cooled, until the final volume is 100 mL.) Thoroughly wet the ash with the acid mixture and return the sample to the muffle furnace. Proceed with the following heating sequence. Heat to 150 °C (over 45 min), to 350 °C (over 1 h), to 550 °C (over 2 h), and, finally, with oxygen passing over the sample, at 550 °C for 10 h or until a white or gray-white ash remains. Cool, as before, to 300 °C. Open the furnace door and continue cooling. Remove the sample when the temperature falls below 150 °C.

When the sample reaches room temperature, add 10 mL 6 M hydrochloric acid, HCl, a couple of glass boiling beads, and 20–30 mL distilled water. Mix the contents of the bottle and heat on a hot plate at 80 °C–90 °C for 30 min. Stir frequently to disperse insoluble silica. Add additional water until the total volume is approximately 50 mL. Afix the bottle cap and agitate the sample upon a mechanical shaker for 15 min. Cool the sample to room temperature and add additional distilled water until the final volume is exactly 55 mL. Allow the dispersed silica to settle by storing the sample for several hours, or overnight. Transfer, from the upper portion of the solution, a measured aliquot for molybdenum determination.

Lyons and Roofayel [28]: The sample is dried for several hours at 105 °C and then ground to a fine powder with a mortar and pestle. Weigh 4 g ground sample into a silica dish. Heat the sample for several hours, or overnight, in a muffle furnace at 550 °C. Remove the sample, cool to room temperature, and add 10 mL distilled water. Follow this with 10 mL 6 M hydrochloric acid, HCl. Cover the sample with a watch glass and gently warm it over a steam bath, swirling the dish to assure the ash is thoroughly wet by the solution. When dissolution appears complete, no more suspended solid seems to dissolve, remove the sample from the steam bath. Filter the sample through a coarse filter paper (e.g. Whatman number 41). Wash the residue thoroughly with distilled water. Collect the filtrate and washings in a 100 mL volumetric flask. Acidify the liquid as required by the procedure to be followed for determining molybdenum and dilute the sample to a final volume of 100 mL with distilled water.

Carel and Wimberley [29] present a method for determining *molybdenum in grasses, soils, and rocks*. Using thiocyanate ion as color forming agent they were, by careful attention to experimental details, able to measure less than 1 µg Mo/mL sample. Colorimetry results agreed favorably with atomic absorption measurements and ion coupled emission spectrometry carried out upon the same samples. With the thiocyanate procedure they found for eight replicate measurements of a grass sample 23.1 µg Mo/g with standard deviation ±0.6.

A *standard molybdenum solution* is prepared by dissolving exactly 0.750 g pure molybdenum(VI) oxide, MoO_3, in 5 mL 0.1 M sodium hydroxide, NaOH. After dissolving carefully acidify the solution by adding 10 mL 1.5 M hydrochloric acid, HCl. Dilute the sample to one liter in a volumetric flask using additional 1.5 M HCl. Exactly 10.00 mL of this solution is placed in a 500 mL volumetric flask and diluted with 1.5 M HCl to its final volume. The diluted solution contains 10.0 μg Mo/mL. Store standard solution in plastic bottles.

Transfer between 0.10 mL (1.0 μg Mo) and 2.00 mL (20.0 μg Mo) to each of several 20 × 150 mm borosilicate culture tubes. Add sufficient 1.5 M HCl to each standard to give a total volume of 10.0 mL in each tube. To a separate culture tube, add 10.0 mL 1.5 M HCl. This is the reference blank.

Place between 1.5 and 2.0 g grass into a weighed 20 × 150 mm culture tube. Dry the sample for several hours, or overnight, in an oven at 70 °C. Cool the sample and reweigh to obtain the sample weight of dried grass. Place the sample tube, mounted vertically, in a cold muffle furnace. Gradually increase the furnace temperature to 500 °C. Maintain this temperature until all organic material is burned away and only ash remains. Remove the sample tube, cool, and add one or two glass beads. Secure the cap on the tube and vigorously shake it to pulverize the sample. Remove the cap and add 1.5 mL 12 M HCl. Replace the cap and put the sample in a steam bath. Heat for one hour. Cool the sample, add 5 mL distilled water, mix, and return the capped tube to the steam bath for one additional hour. Cool the sample and filter, with suction, through a glass fiber filter (e.g. Whatman number 934AH) into a clean 20 × 150 mm culture tube. Wash the filter with 2 to 3 mL distilled water and collect the wash liquid with the sample. Add additional distilled water until the final volume in the culture tube is 10.0 mL.

To each blank, standard, and sample tube, add a small scoop, about 50 mg, sodium tartrate, $Na_2C_4H_4O_6 \cdot 2 H_2O$. Shake each tube to dissolve the tartrate. Add to each tube 0.1 mL 1.3% (w/v) iron(II) ammonium sulfate solution. (Prepare by dissolving 1.3 g $FeSO_4 \cdot (NH_4)_2SO_4 \cdot 6 H_2O$ in 50 mL distilled water. Add 1 mL 18 M sulfuric acid, H_2SO_4, and dilute the solution to 100 mL with distilled water.) Shake each tube. Add to each tube 0.5 mL 10% (w/v) potassium thiocyanate solution. (Prepare by dissolving 10 g KSCN in distilled water and diluting with additional distilled water to a final volume of 100 mL.) Cap each tube, shake, and allow the samples to set for 15 min. Add 2 mL 10% (w/v) tin(II) chloride solution to each tube. (Prepare by dissolving 100 g $SnCl_2 \cdot 2 H_2O$ in 125 mL 12 M HCl, heating if necessary. When dissolved, dilute the solution to 1 liter with distilled water.) Mix each tube and allow it to set for 15 min. Add to each tube 4.0 mL 1:1 (v/v) n-butanol-toluene solution. (Prepare by passing C_4H_8OH through a column packed with silica gel, Davison grade 12, 28–200 mesh. Also, pass $C_6H_5CH_3$ through a separate column packed with the same type silica gel. Mix equal volumes of each liquid.) Cap each tube and shake vigorously for one minute to extract the molybdenum thiocyanate complex. Allow a few minutes for each tube to separate into distinct layers. Withdraw the organic layer and filter it through a course filter paper (e.g. Whatman number 41) into a small glass vial. Cap each vial until ready to measure absorbance. Record the absorbance at 465 nm using 1 cm cells. Measure each standard and sample against the zero μg Mo solution as reference. Plot the data for the standards, that is absorbance readings versus amount of molybdenum present in each standard sample. From this curve read, for the unknown sample, the molybdenum content corresponding to its observed absorbance reading. Correct this molybdenum value for the sample size and report μg Mo/g dry sample.

3. Soil Samples

Molybdenum content of soils in which plants are grown has received considerable attention [30–32]. Present from trace amounts to several μg Mo/g soil, molybdenum distribution varies with type of soil and climate. One must distinguish between available molybdenum, those forms of molybdenum readily usable by plants, and total molybdenum, the total amount of molybdenum present, in all forms. Various attempts have been made to relate molybdenum content in plants and in animals grazing upon these plants to available molybdenum in soil [33–36]. The attempts are not always successful. Whether available molybdenum or total molybdenum content of soil is measured depends largely upon the method of sample preparation. Leaching of soluble molybdate from soil by prolonged agation in the presence of water [37], ammonium acetate [38], ammonium oxalate [39,40], or mixed organic acids [41] is possible and measures available molybdenum. Oxalate extraction [39] is perhaps the best known of these procedures and serves as an example by which the other methods are compared. Leaching with mineral acids results in molybdenum values not related to plant molybdenum content [42]. Total molybdenum is obtained by wet oxidation [43,44], dry decomposition [45], or sodium carbonate fusion [46,47] of soil samples.

Manzoori [38] – *available Mo*: Place 50 g soil sample in a one liter beaker. Add 800 mL 1 M ammonium acetate solution, $NH_4C_2H_3O_2$. The mixture is covered and stirred for several hours, or overnight. Filter the suspension through an 18.5 cm medium speed filter paper (e.g. Whatman number 540). Wash the paper and residue several times with distilled water. Collect filtrate and washings in a one liter beaker and evaporate the solution to near dryness on a hot plate. Transfer the beaker to a steam bath and continue evaporating until the solvent is completely gone. Thoroughly rinse the beaker with distilled water, dissolve the residue, and place the sample solution in a 100 mL volumetric flask. Add additional distilled water, or acid depending upon the procedure to be followed for molybdenum, until the final solution is 100 mL.

Kim, Owens, and Symthe [44] – *total Mo*: A weighed one gram soil sample, previously air dried and ground to 100 mesh or finer, is placed in a 100 mL teflon beaker. Add 4 mL 70% (w/v) perchloric acid, $HClO_4$, and 10 mL 40% (w/v) hydrofluoric acid, HF. Swirl the mixture to wet the sample with the acids and place the beaker on a hot plate. Continue heating until it appears nor further reaction will occur. Evaporate the sample until fumes from perchloric acid are observed. Use caution when fuming perchloric acid solutions. A special fume hood is required to minimize the danger of explosion. Add 10 mL 40% (w/v) HF, swirl the mixture and again evaporate the solution until fumes from perchloric acid appear. Wash the sides of the beaker with 1 M hydrochloric acid, HCl. Add additional 1 M HCl until the final volume is about 25 mL. Cover the beaker with a watch glass and boil the sample for several minutes, until the residue dissolves. Cool the sample and transfer it to a 50 mL volumetric flask. Wash the beaker with 1 M HCl and add the washing to the flask. Dilute the sample to exactly 50 mL with additional 1 M HCl. Specific laboratory directions for atomic absorption measurement of molybdenum following this sample treatment are given in chapter 8.

Hesse [47] – *total Mo*: A weighed 1 g sample of soil, previously dried at 105 °C in an oven for one hour to remove moisture, is mixed with 6 g anhydrous sodium carbonate, Na_2CO_3. Place the mixture in a platinum crucible. Add additional Na_2CO_3 to form a cover layer over the top of the mixture. Place the crucible, partially covered, in a muffle furnace and heat to 1,200 °C. When the sample is liquid, briefly remove the crucible with platimun tipped tongs (to avoid contamination by other metals) and swirl the crucible forming a thin layer of melt around the inner crucible wall. Return the crucible to the furnace for 15 min additional heating. Remove the sample and again swirl the content to cover the sides of the platimun crucible. Place the sample in a 400 mL beaker and when cool ad sufficient distilled water to cover the crucible. Heat the beaker on a steam bath until the melt separates from the crucible. Using tongs, remove the crucible and lid. Rinse each with distilled water catching the water in the 400 mL beaker. Place the crucible and lid in a 250 mL beaker. Partially cover the crucible with 5 M hydrochloric acid, HCl, containing 1% (v/v) ethanol, C_2H_5OH. Heat the crucible on a hot plate. This assures that all of the melt has been removed from the crucible. Carefully add, to the sample solution in the 400 mL beaker, 10 mL 12 M HCl and 10 mL 70% (w/v) perchloric acid, $HClO_4$. When evolution of carbon dioxide, CO_2, ceases, add the HCl rinse solution from the 250 mL beaker containing the platinum crucible and lid. Rinse the crucible and cover with distilled water and add this to the bulk sample solution. Cover the sample solution with a raised watch glass and evaporate the sample on a sand bath. Use caution in evaporating to dryness samples containing $HClO_4$. After the sample is dry, continue baking it for 15 min to convert hydrous silica present to insoluble silicon dioxide, SiO_2. Dissolve the residue in 25 mL warm water. Filter the sample through a course filter paper (e.g. Whatman number 41). Rinse the residue on the paper with 0.5 M HCl. Collect filtrate and rinsings in a 100 mL volumetric flask and dilute the sample to exactly 100 mL with additional 0.5 M HCl.

Carel and Wimberly [29], in addition to measuring the molybdenum content of plant samples, apply their colorimetric thiocyanate procedure to determination of molybdenum in soils.

Consult the preceeding section for preparation of molybdenum standard solutions.

For *soil samples* weigh 0.3 g air dried, finely ground sample and place it in a 20 × 150 mm borosilicate culture tube. The sample should not contain more than about 18 µg Mo. Add 1.5 g potassium pyrosulfate, $K_2S_2O_7$. Thoroughly mix the dry materials. Place the tube, vertically mounted, in a cold muffle furnace. Gradually increase the temperature of the furnace to 500 °C. Maintain this temperature for two hours. Remove and cool the sample. Add 1.25 mL 12 M hydrochloric acid, HCl. Cap the tube and place it in a steam bath for one hour. When cool, add 8.75 mL distilled water. Replace the cap and return the sample to the steam bath. Continue heating for one hour shaking the tube initially to dissolve the melt. Remove the tube and cool it to room temperature. Filtration of the sample at this point is unnecessary.

1.5 g $K_2S_2O_7$ is placed in each of several 20 × 150 mm borosilicate culture tubes and fused along with the soil sample. Upon removal from the furnace and cooling, aliquots of standard molybdenum solution are added to each tube. Add 0, 0.10 mL (1.0 µg Mo) and up to 2.00 mL (20.0 µg Mo) of standard solution. To each of these tubes, add 1.25 mL 12 M HCl. Carry through with heating on the steam bath as described in the preceeding paragraph. After the first heating, add sufficient distilled water to each tube until its final volume is 10.0 mL.

Blank, standards, and sample are now treated as described in the preceeding section beginning with the addition of solid sodium tartrate.

References

1. Bowen, H.J.M.: Trace Elements in Biochemistry 1966 New York, Academic Press
2. Tompsett, S.L., Fitzpatrick, J.: Analyst (London), *75*, 279 (1950)
3. Versieck, J., Cornelis, R.: Anal. Chim. Acta, *116*, 217 (1980)
4. Christian, G.D., Patriarche, G.J.: Anal. Lett., *12B*, 11 (1979)
5. Bentley, G.E., Markowitz, L., Meglen, R.R.: Ultratrace Metal Analysis in Biological Science and Environment, Advancements in Chemistry Series, Number 172 (Risby, T.H. ed.) 1979, Washington, D.C., Amer. Chem. Soc.
6. Swedo, K.B., Enemark, J.H.: J. Chem. Educ., *56*, 70 (1979)
7. Coughlan, M.P.: Molybdenum and Molybdenum Containing Enzymes, 1980, Oxford, Pergamon Press

8. Sittig, M.: Toxic Metals, 1976, Park Ridge, NJ, Noyes Data Corporation
9. Luckey, T.D., Venugopal, B.: Metal Toxicity in Mammals, 1977, New York, Plenum Press
10. Agrawal, Y.K., Anal. Lett., *13B*, 357 (1980)
11. Verstuyff, A.W.: Analytical Techniques in Occupational Health, Chemistry Symp. Ser. Number 120 (Dollberg, D.D., Verstuyff, A.W. eds) 1980, Washington, D.C., Amer. Chem. Soc.
12. Pijck, J.: Verh. K. Vlaam. Akad. Wet. Belg. Kl. Wet. Number 67, (1961); Chem. Abstr., *57*, 2531d (1962)
13. Morsches, B., Tölg, G.: Fresenius' Z. Anal. Chem., *219*, 61 (1966)
14. Norheim, G., Waasjø, E.: Fresenius' Z. Anal. Chem., *286*, 279 (1977)
15. Postgate, J.R.: The Chemistry and Biochemistry of Nitrogen Fixation, 1971, London, Plenum Press
16. Brey, R.C., Swann, J.C.: Struct. Bonding (Berlin), *11*, 107 (1972)
17. Chem. Eng. News, *51*(39), 15 (1973)
18. Zumft, W.G.: Struct. Bonding (Berlin), *29*, 41 (1976)
19. Chappell, W.R., Petersen, K.K.: Molybdenum in the Environment, 1976, New York, Marcel Dekker
20. Newton, W.E., Otsuka, S.: Molybdenum Chemistry of Biological Significance, 1980, New York, Plenum Press
21. Adhate, S.: Indian J. Agron., *5*, 57 (1960)
22. Sauchelli, V.: Trace Elements in Agriculture, 1969, Princeton, New Jersey, Van Nostrand-Reinhold
23. Nichols, M.L., Rogers, L.H.: Ind. Eng. Chem., Anal. Ed., *16*, 137 (1944)
24. Benne, E.J., Jerrim, D.M.: J. Assoc. Off. Agric. Chem., *39*, 412 (1956)
25. Smit, J., Smit, J.A.: Anal. Chim. Acta, *8*, 274 (1953)
26. Benne, E.J., Linden, E.I.: J. Assoc. Off. Agric. Chem., *43*, 510 (1960)
27. Heanes, D.L.: Analyst (London), *106*, 172 (1981)
28. Lyons, D.J., Roofayel, R.L.: Analyst (London), *107*, 331 (1982)
29. Carel, A.B., Wimberley, J.W.: Anal. Lett. *15A*, 493 (1982)
30. Sequi, P.: Agrochimica, *9*, 260 (1972), Chem. Abstr., *78*, 83159t (1973)
31. Aubert, H., Pinta, M.: Trace Elements in Soils, 1977, Amsterdam, Elsevier
32. Davies, B.E.: Applied Soil Trace Elements, 1980, New York, John Wiley
33. Purvis, E.R., Peterson, N.K.: Soil Sci., *81*, 23 (1956)
34. Haley, L.E., Melsted, S.W.: Soil Sci. Soc. Amer., Proc., *21*, 316 (1957)
35. Mekenzie, R.: Aust. J. Exp. Agric. Anim. Husb., *6*, 170 (1966)
36. Healy, W.B., Bate, L.C., Ludwig, T.G.: N. Z. J. Agric. Res., *7*, 603 (1964), Chem. Abstr., *62*, 8189h (1965)
37. Lowe, R.H., Massey, H.F.: Soil Sci., *100*, 238 (1965)
38. Manzoori, J.L.: Talanta, *27*, 682 (1980)
39. Grigg, J.L.: Analyst (London), *78*, 470 (1953)
40. Gorlach, E.: Rocz. Glebozn., *14*, 15 (1964), Chem. Abstr., *64*, 1344c (1966)
41. Golubev, I.M.: Agrokhimiya, 135 (1966), Chem. Abstr., *65*, 6233g (1966)
42. Ivanov, D.N., Bol'shakov, V.A.: Khim. Sel'sk. Khoz., *7*, 229 (1969), Chem. Abstr., *71*, 29593x (1969)
43. Scharrer, K., Munk, H.: Agrochimica, *1*, 44 (1956)
44. Kim, C.H., Owens, C.M., Smyth, L.E.: Talanta, *21*, 445 (1974)
45. Savinova, E.N., Korobova, E.M., Shumskaya, T.V.: Zh. Anal. Khim., *36*(7), 1267 (1981)
46. Robinson, W.O.: J. Assoc. Off. Agric. Chem., *38*, 246 (1955)
47. Hesse, P.R.: A Textbook of Soil Chemical Analysis, 1971, London, John Murray Publisher

Chapter 19

Environmental Samples

Man's increased awareness of his environment has necessitated development of sensitive analytical techniques to detect and determine traces of material in natural samples.

1. Water Samples

Water testing, rivers, lakes, seas, oceans, ground water, waste water, etc. is now routine for both organic and inorganic components using a variety of published methods [1–3]. Although present to only a small extent in natural waters, some attention has been directed specifically towards molybdenum [4, 5]. Because of the minute amounts of molybdenum involved, on the order of µg Mo/L or less, preconcentration generally preceeds quantitative measurement of molybdenum by whatever methods employed. Coprecipitation, extraction, ion exchange, and other techniques are used to collect molybdenum from large sample volumes into smaller, more concentrated solutions. Frequently, too, removal of possible interferences is accomplished at the same time. These procedures are discussed in more detail in chapter 3. Sample storage, for trace analysis, is also a concern. Adsorption from the sample onto the container surface, glass or plastic, can remove a relatively large fraction of the metal ions present. Leaching of impurities present either upon or within the container wall can, on the other hand, result in a relatively large fractional increase of the sought for constituent. Reagent purity, too, must be considered when performing trace analyses. As part of his study on pH effects of long term stability of metal ion solutions Smith [6] recommends molybdate samples be kept in solution at pH 1.5. Two studies not previously discussed for determination of molybdenum in natural waters may be useful if instrumentation is available. They measure molybdenum by electron paramagnetic resonance [7] and spark source mass spectrometry [8].

Kim and Zeitlin [9] – *sea water*: 500 mL of sample is placed in a 1 L beaker and acidified with 1 mL 9 M sulfuric acid, H_2SO_4. Add 3.0 mL 0.1 M thorium nitrate, $Th(NO_3)_4 \cdot 4H_2O$. Add dropwise with stirring 15 M ammonia, $NH_3 \cdot H_2O$, solution. Continue adding ammonia until the pH is 6.0, as read from a pH meter. Allow the solution to set for 30 min before collecting the precipitate on a 0.8 µm membrane filter (e.g. Gelman number AN-800). Discard the filtrate. Dissolve the residue from the filter with 2–3 mL 12 M hydrochloric acid, HCl, collecting the solution in a 50 mL beaker. Evaporate the sample to dryness on a low temperature hot plate. After the dry residue cools to room temperature, dissolve the sample in 5 mL 6 M HCl. Continue with the determination of molybdenum.

Zaguzin, Ksenzova, and Pogrebnyak [10] determine tungsten, molybdenum, and tin in natural waters by emission spectroscopy after first concentrating these elements on an activated carbon-tannin adsorbent following sample adjustment to pH 2. Silicon, aluminum, and iron also precipitate under these conditions but do not seem to interfere with the subsequent spectrographic determination. After adsorbent addition samples are filtered to collect the solid, which is then ashed and packed in a graphite electrode for dc arc excitation. Molybdenum is measured in the range 0.03–10.0 µg Mo/L using the 281.615 nm emission line and from 10.0–300 µg Mo/L using the less sensitive 264.946 nm emission. Standard deviation values reported by the authors for replicate measurements of synthetic samples at various molybdenum concentrations are:

Mo concentration, µg/mL	standard deviation
0.03	± 0.3
0.1	± 0.15
1.0	± 0.1
10.0	± 0.08
100.0	± 0.05

The authors suggest that samples stored for several days prior to measurement be adjusted to pH 10 with sodium hydroxide to minimize adsorption of metal ions, especially tungsten, on the container surface. Glass storage containers are recommended.

Operating Conditions

Electrodes	sample (anode)	graphite rod 5 mm outside diameter, drilled 3 mm deep by 3 mm diameter
	counter (cathode)	graphite rod shaped to a point
Exication	dc arc, 13–14 amp	
Spectrograph	model DFS-13, grating 1200 lines/mm first order	
Slit width	27 µm	
Exposure time	30–35 sec	
Photographic plate	Kodak spectroscopic plate type II	

A *standard molybdenum solution* is prepared by dissolving 0.630 g sodium molybdate, $Na_2MoO_4 \cdot 2H_2O$, in water. Dilute the sample to exactly one liter in a volumetric flask. Transfer 20.0 mL of this solution to a 500 mL volumetric flask. Dilute this sample with distilled water to 500 mL. This solution, solution A, contains 10.0 µg Mo/mL. Transfer 10.0 mL solution A to a 1,000 mL volumetric flask. Add distilled water until the volume of this flask is 1,000 mL. This solution, solution B, contains 0.10 µg Mo/mL.

The salt content of the prepared water samples containing known amounts of molybdenum is approximated by dissolving 0.355 g calcium chloride, $CaCl_2 \cdot 2H_2O$, per liter of sample. Samples are adjusted to pH 2 with dropwise addition of 6 M hydrochloric acid, HCl. Use a pH meter. Should closer approximation to a natural water sample be desired, additional salts will have to be added to the sample [11]. Add, for the 0.03–10.0 µg Mo/L series, varying amounts of molybdenum stock solution B to separate one liter beakers. Use from 0.30 mL (0.03 µg Mo) to 100 mL (10.0 µg Mo) solution B. Dilute each sample to exactly one liter with the calcium chloride, synthetic water, solution. For the 10.0–300 µg Mo/mL series add various amounts of molybdenum stock solution A to separate one liter beakers. Use from 1.00 mL (10.0 µg Mo) to 30.0 mL (300 µg Mo) solution A. Dilute each sample to exactly one liter with the calcium chloride solution. If tungsten and tin are to be determined along with molybdenum, appropriate stock solutions containing these elements should also be present in the standard solutions. One liter of calcium chloride solution, at pH 2, but containing no molybdenum, serves as the reference blank sample.

One liter of water sample is adjusted to pH 2 by dropwise addition of 6 M HCl.

Add to each blank, standard, and sample 0.5 g 1:1 (w/w) mixture of activated carbon-tannin. (Prepare activated carbon by heating at 110 °C for one hour just prior to use. Cool in a desiccator weigh, and mix with an equal weight tannin. Use at once.) Stir each mixture with a magnetic stirrer for approximately one-half hour. Allow the solids to settle and filter, in turn, each blank, standard, and sample through separate fine filter paper (e.g. Schleicher & Schuell number 589 blue ribbon). Wash each residue with 10 mL 0.01 M HCl. Each filter paper is folded and transferred to a porcelain crucible. Place each crucible in a drying oven at 70 °C to remove moisture. When dry, place each crucible in the low heat of a burner to char the paper. When the paper is sufficiently burned away, remove each crucible from the flame and place them in a muffle furnace. Gradually increase the furnace temperature to 550 °C. Maintain this temperature for one hour. Inspect the ash, if white or gray-white, heating has been sufficient; if black carbon deposits remain, continue heating in the furnace. When ashing is complete, remove the samples and cool them to room temperature. The authors did not use a spectroscopic buffer but simply packed the electrode cavity directly with powdered sample. One may wish to consider mixing samples with an appropriate buffer (see Table 7-2) prior to excitation. Each blank, standard, and unknown is excited for 30–35 s at 13–14 A using an electrode gap of 3 mm.

Exposed plates are developed and emission line intensities obtained by the usual spectroscopic procedures. Those unfamiliar with these procedures may wish to consult an appropriate reference [12, 13]. Background correction for the emission lines is read from the plate in a transparent region near the analytical line. The authors used no internal standard in their measurements. The amount of molybdenum present in the one liter sample is read from the calibration plots based upon the emissions from the known solutions.

2. Air Samples

Air like water is a necessary resource and testing of air for dissolved gases, other soluble materials, and particulates is common [14–18]. Molybdenum is generally present in amounts less than 1 ng/m^3 air [15] although higher concentrations are observed near mining sites and ore processing facilities. Studies specific for molybdenum in air are cited in the review literature [19, 20] and comprehensive studies of city areas [21], rural areas [22], areas near atomic testing sites [23], and areas near welding operations [24] all include molybdenum as one of the components present.

Imai, Kusaka, Tsuji, and Hishiya [25] – *insoluble Mo*: As part of a study on the presence of 14 trace metals, including molybdenum, the authors collected dust samples by passing air through a membrane filter (Gelman number GA-3, cellulose acetate filter, 1.2 µm pore size, 110 mm diameter). Commercial air sampling equipment was used and the supported filter suspended at the collection site 10 m above the ground. Flow rate was 20 L/min and sampling continuous for one month. To remove trace metal impurities from the membrane filter prior to use, immerse it in 2 M nitric acid, HNO$_3$, for 6 h. Remove the filter, wash it thoroughly with double distilled water, and place it in 2 M acetic acid, $HC_2H_3O_2$, for 6 h. Remove the filter, wash it thoroughly with double distilled water, and dry it in a desiccator over silica gel for 24 h. Weigh the dried filter. After samples are collected, dry the filter and sample in a desiccator. Weigh and repeat the drying and weighing until constant weight is obtained. Weight of the particulates retained on the filter is found by difference. The filter with sample is placed in a quartz boat and put in a commercial low temperature asher for 2 h. The authors proceed to mix the dry ash with spectroscopic buffer in preparation for emission analysis. An alternate approach would be to dissolve the ash in acid and proceed with some other type of measurement.

3. Fuels Samples

There is a growing awareness of the need to conserve natural fuel resources, namely coal and oil. Much attention is being directed towards the composition of these materials including measurement of trace metal content [26–31]. One recent study [29] discusses methods of sample preparation, high and low temperature ashing, fusion, wet decomposition, etc. of coal. In addition they report values for over 60 elements present in National Bureau of Standards (USA) coal and fly ash reference materials. Determination of trace metals in petroleum products, generally refined petroleum products, gasoline, fuel oils, etc. is also discussed [31]. Molybdenum, specifically, is determined in shale oil [32].

Nadkarni [29] – *coal*: Place 1–2 g finely powdered coal in a weighed borosilicate petri dish and put it in a commercial low temperature asher. Every 2–3 h remove the dish, stir the sample to expose a fresh surface, and weigh the dish and its contents. Repeat this procedure until constant sample weight is obtained. This may take up to 24 h. The dry ash is dissolved according to the procedure selected for molybdenum determination.

Hofstader, Milner, and Rummels [31] – *refined petroleum fuel*: Weigh between 20 and 100 g sample, depending upon the suspected molybdenum content, and place it in a vycor dish. Slowly and carefully add 5 mL 18 M sulfuric acid, H_2SO_4. Mix the contents of the dish after adding each increment of acid. For volatile fuels, evaporate as much of the sample as possible by directing a stream of nitrogen across the sample. Suspend the dish in a large beaker, air bath, and place the beaker on a hot plate but do not heat the sample yet. Suspend an infrared heat lamp approximately 3 cm above the dish. Heat the sample with the lamp and when no further evaporation occurs, gradually increase the temperature of the hot plate. Stir the sample frequently during heating to avoid spatter. Eventually remove the lamp and increase the temperature of the hot plate. Continue heating the sample until fumes of sulfur trioxide, SO_3, are observed. Bake the sample until all SO_3 is removed. Transfer the dish to a muffle furnace at 480 °C. Direct a stream of oxygen across the sample while in the furnace. Continue heating until a white or gray-white ash remains, all carbon having been burned away. Remove the sample from the furnace and cool to room temperature. Carefully add 5 mL 6 M hydrochloric acid, HCl, to the sample directing the acid around the sides of the dish to assure all particles of ash are dissolved. When dissolved, quantitatively transfer the sample to a 10 mL volumetric flask and dilute to a final volume of 10 mL with additional 6 M HCl.

Saba and Eisentraut [33] – *used lubricating oil*: This procedure is intended for preparing samples for aspiration into a flame atomic absorption spectrometer. Five gram weighed sample is placed in a 2 oz polyethylene bottle. Carefully add 1.25 g nitric acid-hydrofluoric acid mixture. (Prepare by combining, by volume, 7 parts 15 M HNO_3 and 1 part 40% (w/v) HF.) Stir the oil-acid mixture. Cap the bottle and shake it vigorously, preferably with a mechanical shaker, for 2 min. Add 31 g (approximately 38 mL) methyl isobutyl ketone. Mix the liquids until a single phase results. The sample is now ready for direct aspiration into an atomic absorption burner.

References

1. Förstner, U., Wittmann, G.T.W.: Metal Pollution in the Aquatic Environment, 1981, Berlin, Springer-Verlag
2. Florence, T.M.: Talanta, 29, 345 (1982)
3. Golterman, H.L., Clymo, R.S., Ohnstad, M.A.M.: Methods for Physical and Chemical Analysis of Fresh Water, 1978, Oxford, Blackwell Scientific Publications
4. Fishman, M.J., Mallory, E.C.: J. Water Pollut. Control Fed., 40, R 67 (1968)
5. Monien, H., Bovenkerk, R., Kringe, K.P., Rath, D.: Fresenius' Z. Anal. Chem., 300, 363 (190)
6. Smith, A.E.: Analyst (London), 98, 65 (1973)
7. Hanson, G., Szabo, A., Chasteen, N.D.: Anal. Chem., 49, 461 (1977)
8. Vanderborght, B.M., Van Grieken, R.E.: Talanta, 27, 417 (1980)
9. Kim, Y.S., Zeitlin, H.: Anal. Chim. Acta, 51, 516 (1970)
10. Zaguzin, V.P., Ksenzova, V.I., Pogrebnyak, Yu.F.: Zh. Anal. Khim., 35(6), 1143 (1980)
11. Danielsson, L.-G., Magnusson, B., Westerlund, S.: Anal. Chim. Acta, 98, 47 (1978)
12. Brode, W.R.: Chemical Spectroscopy 1943 New York, John Wiley & Sons
13. Grove, E.L.: Applied Atomic Spectroscopy, 1978, New York, Plenum Press
14. Ledbetter, J.O.: Air Pollution, 1972, New York, Marcel Dekker
15. Warner, P.O.: Analysis of Air Pollutants, 1976, New York, John Wiley
16. Perry, R., Young, R.J.: Handbook of Air Pollution Analysis, 1977, London, Chapman and Hall
17. Butler, J.D.: Air Pollution Chemistry, 1979, London, Academic Press
18. Dams, R., Billiet, J., Block, C., Demuynck, M., Janssens, M.: Atm. Environ., 9, 1099 (1975)
19. Vengerskaya, Kh.Ya.: Nov. Obl. Prom.-Sanit. Khim. (Murav'eva, S.I. ed) 1969, Moscow, Izd. Meditsina; Chem. Abstr., 71, 116274n (1969)
20. Tabor, E.C.: Health Lab. Sci., 7, 149 (1970)
21. Hettche, H.O.: Air Water Pollut, 8, 185 (1964)
22. McMullen, T.B., Faoro, R.B., Morgan, G.B.: J. Air Pollut. Control Assoc., 20, 369 (1970)
23. Benson, P., Gleit, C.E., Leventhal, L.: US AEC Symp. Ser. 5, 108 (1965); Chem. Abstr., 66, 7745b (1967)
24. Kosternaya, A.F., Zavorovskaya, N.A.: Vses. Nauchno-Issled. Inst. Okhr. Tr., 297 (1967); Chem. Abstr., 70, 92904u (1969)
25. Imai, S., Kusaka, Y., Tsuji, H., Hishiya, Y.: Anal. Chim. Acta, 108, 103 (1979)
26. Radmacher, W., Hersling, H.: Fresenius' Z. Anal. Chem., 167, 172 (1959)
27. Butsik, L.A., Derbaremdiker, M.M., Korolev, M.M.: Vop. Geol. Sev.-Zap. Sekt, Tikhookean. Poyasa, Vladivostok, 166 (1966); Chem. Abstr., 66, 97531x (1967)
28. Ratynskii, V.M., Glushnev, S.V.: Dokl. Akad. Nauk SSSR, 177, 1193 (1967)
29. Nadkarni, R.A.: Anal. Chem., 52, 929 (1980)
30. Colombo, U., Sironi, G.: Anal. Chem., 36, 802 (1964)
31. Hofstader, R.A., Milner, O.I., Runnels, J.H.: Analysis of Petroleum for Trace Metals, 1976, Advances in Chemistry Series 156, Washington, D.C., Amer. Chem. Soc.
32. Vahemets, H.: Uch. Zap. Tartu Gos. Univ. No.193, 121 (1966); Chem. Abstr., 69, 40971y (1968)
33. Saba, C.S., Eisentraut, K.J.: Anal. Chem., 51, 1927 (1979)

Chapter 20

Miscellaneous Samples

Because of its importance in plant growth, molybdenum is sometimes added to fertilizers, generally in amounts of 100 µg Mo/g or less. One study of its determination in *fertilizers* by atomic absorption measurement produced rather different results from different laboratories especially with low levels of molybdenum [1].

Koirtyohann and Hamilton [2] – *fertilizer*: Add to one gram fertilizer sample in a 250 mL beaker 20 mL 6 M hydrochloric acid, HCl. Heat the sample on a hot plate for 30 min. Continue heating until the sample volume is reduced to about 10 mL. Add 50 mL distilled water and increase heating until the sample boils. Cool the sample to room temperature and filter through a course filter paper (e.g. Whatman number 541). Rinse the beaker and filter residue with 4–5 mL distilled water and then with 2–3 mL 8 M ammonia, $NH_3 \cdot H_2O$. Repeat the ammonia rinse with an additional 2–3 mL. Finally, rinse the residue with 4–5 mL distilled water. All rinsings are collected along with the filtrate in a 100 mL volumetric flask. If the solution is basic, add sufficient 6 M HCl until the sample becomes acid. Add distilled water until the sample is exactly 100 mL. Proceed with molybdenum determination.

An important use of molybdenum is as a *catalyst* in various industrial processes. Perhaps the most widespread use is for removal of sulfur, hydrodesulfurization, in petroleum and coal [3, 4]. Molybdenum(VI) oxides in combination with cobalt or nickel and on an aluminum oxide, Al_2O_3, support effectively converts organic sulfur compounds present in petroleum and/or coal to hydrogen sulfide, resulting in a more desirable, reduced sulfur fuel. Along with MoO_3, molybdenum(IV) sulfide, molybdoheteropoly species, and other molybdenum compounds catalyze a variety of processes. These include oxidation of naphthalene to phthalic anhydride and disproportionation of olefins [5]. Both electron paramagnetic resonance spectrometry [6] and x-ray photoelectron spectrometry [7] are applied to analysis of molybdenum containing catalysts.

Labrecque [8] outlines an atomic absorbance procedure for determination of molybdenum in *hydrodesulfurization catalysts*. Cobalt and/or nickel are measured in the same sample. Aluminum, present in the catalyst, enhances the observed signal for molybdenum so it is necessary to add aluminum ion to the molybdenum standards. The alternative to this is to use a standard addition procedure. Samples are treated with sulfuric acid and hydrofluoric acid added to remove silica. A correction for adsorbed moisture is necessary and is obtained from weight loss on ignition of a separate catalyst sample. For molybdenum and cobalt, nitrous oxide-acetylene flame is used. Nickel is measured with an air-acetylene flame. For molybdenum monitor the 313.3 nm emission using a 5 mA lamp current. Slit width is 0.10 mm and aspiration rate 3.1 mL/min. Optimum instrument settings for cobalt and nickel are listed in the original paper. For an actual hydrodesulfurization catalyst containing Mo–Co–Ni, the author found from six repetitive measurements a mean value and standard deviation for MoO_3 of $(14.8 \pm 0.3)\%$. The same sample contained $(1.28 \pm 0.02)\%$ Co and $(3.09 \pm 0.03)\%$ Ni.

To prepare a *standard molybdenum solution*, weigh 0.750 molybdenum(VI) oxide, MoO_3, and place it in a 100 mL beaker. Add 50 mL 10% (w/v) sodium hydroxide, NaOH. Dissolve the sample. Slowly add 20 mL 9 M sulfuric acid, H_2SO_4. Cool the solution and quantitatively transfer it to a 500 mL volumetric flask. Add additional distilled water to the flask until the final volume is exactly 500 mL. The concentration of this solution is 1.00 mg Mo/mL. Place, in separate 100 mL volumetric flasks zero, 1.00, 2.00, 5.00, 7.00, and 10.00 mL of this stock solution. When diluted to their final volumes, these solutions will contain 0.00, 10.0, 20.0, 50.0, 70.0, and 100.0 µg Mo/mL respectively.

A weighed 0.5 g catalyst sample is placed in a 125 mL teflon beaker. Add 20 mL 18 M H_2SO_4 followed by careful addition of 10 drops 40% (w/v) hydrofluoric acid, HF. After the reaction with HF subsides, add 5.0 mL 40% (w/v) HF. Heat the covered beaker on a hot plate at 90 °C for 45 min. If undissolved silica remains, cloudy solution, add an additional 10.0 mL 40% (w/v) HF and continue heating until a clear solution results. When clear, maintain the uncovered beaker at 60–80 °C for 1–2 h to evaporate excess HF. Add 20 mL distilled water and continue heating at 80 °C for 20–30 min. Pour the hot solution into a 100 mL volumetric flask and, when cool, add distilled water to make a final volume of 100 mL.

Transfer exactly 1.00 mL of unknown solution to each of the flasks containing the standard molybdenum aliquots. Also add 1.00 mL unknown solution to the blank. Dilute each standard and the blank to exactly 100 mL with distilled water. Shake each flask briefly to mix its contents. Aspirate the most concentrated molybdenum standard into the flame. Acetylene flow is 4 L/min and N_2O flow is 9 L/min. Carefully fine adjust each flow rate to achieve maximum absorbance reading from the sample. At the same time adjust burner height for maximum reading. After final instrument settings are made, aspirate, in turn, each sample into the flame. Record absorbance values. Plot absorbance values versus concentration of standard molybdenum, µg Mo/mL, for each standard and the blank. The standard addition curve obtained is displaced upwards from the origin. Extend the curve to the left of zero until it reaches the concentration axis. The molybdenum value at this intersection is the concentration of molybdenum in the unknown solution. From this value, calculate the total amount of molybdenum present in the sample and, using the dry sample weight, the percent Mo in the catalyst. One may wish to express this percent in terms of MoO_3 rather than Mo.

Another important use of molybdenum, in the form of molybdenum(IV) sulfide, MoS_2, is as a *lubricant*. Possessing a structure similar to graphite, MoS_2 serves well as a solid lubricant and as an additive to various lubricating oils [9, 10]. One method for analyzing solid lubricants is to burn the sample in a stream of oxygen. Sulfur dioxide, SO_2, and molybdenum(VI) oxide, MoO_3, are liberated, collected separately, and weighed [11]. Other procedures for determining molybdenum in lubricants are cited in various earlier chapters.

Julietti and Wilkinson [12] report an atomic absorption procedure for determination of molybdenum in solid molybdenum(IV) sulfide lubricants. These lubricants generally contain, in addition to MoS_2, graphite and a resin binder.

The sample is fused with sodium hydroxide to render molybdenum soluble while at the same time burning away the graphite and resin. No loss of molybdenum is observed if the fusion is carried out with a flame and then, after molybdenum has reacted with the flux, temperature increased in a muffle furnace to remove carbon materials. Sulfuric acid dissolves the melt and is also added to the molybdenum standards thus presenting a more nearly identical matrix for sample and standards. Measurement is made with air-acetylene using a 5 cm slot burner. The 313.3 nm molybdenum line is monitored. Other instrument settings include a 5 mA lamp current, 0.05 mm slit width setting, and 2 mL/min sample aspiration rate. The authors report a mean result from the atomic absorption procedure within $\pm 0.1\%$ of the mean value for identical samples obtained by gravimetry. Samples contained approximately 8% molybdenum as Mo.

Molybdenum(VI) oxide, MoO_3, is dried for one hour in an oven at 150 °C and cooled to room temperature. Weigh 0.600 g MoO_3 and place it in a 100 mL beaker. Add 50 mL 10% (w/v) sodium hydroxide, NaOH. Dissolve the sample. Now slowly add 20.0 mL 9 M sulfuric acid, H_2SO_4. Cool the solution and quantitatively transfer it to a 250 mL volumetric flask. Add additional distilled water to the flask until the final volume is exactly 250 mL. The concentration of this solution is 1.60 mg Mo/mL. For best results it is important that standards and sample contain approximately the same sulfate concentration. Therefore, to prepare each standard transfer, to separate 100 mL volumetric flasks, the following volume of standard molybdenum solution and 1.0 M sulfuric acid, H_2SO_4.

volume Mo standard, mL	volume 1.0 M H_2SO_4, mL	total Mo present, µg/mL
0.00	7.2	0.0
1.00	6.5	16.0
2.00	5.8	32.0
5.00	3.6	80.0
7.00	2.2	112
10.00	0	160

Add distilled water to each flask until the final volume of each is exactly 100 mL. Shake each flask to mix its contents.

Place 5 g sodium hydroxide, NaOH, pellets in a nickel crucible of approximately 5 cm diameter. Fuse the NaOH over a burner and, upon cooling, rotate the curcible to coat the sides with molten NaOH. Weigh a 0.5 g sample of lubricant and distribute it evenly around the inside of the crucible. Fuse the sample, without cover, over an open flame until effervescence subsides, about 15 min. Place the sample in a muffle furnace at 650 °C for 30 min. Remove the crucible, cool to room temperature, and place it upright in 50 mL distilled water in a 250 mL beaker. Carefully add 20.0 mL 9 M H_2SO_4. Dissolve the melt from the crucible heating if necessary to dislodge the solid. Remove the crucible while directing a stream of distilled water upon it for rinsing. Collect the washing in the beaker along with the dissolved sample. Boil the solution for 5 min to remove any hydrogen sulfide which may be present. Filter the cool solution through a medium filter paper (e.g. Whatman number 40) into a 250 mL volumetric flask. Wash the paper briefly with distilled water and collect the washing in the flask. Dilute the sample to exactly 250 mL with additional distilled water.

With air pressure at 15 lb/in², aspirate a portion of the most concentrated standard into the air-acetylene flame and adjust acetylene flow to obtain maximum absorbance reading. At the same time, adjust burner height for maximum reading. When this is done, aspirate, in turn, each blank, standard, and unknown sample into the flame. Record absorbance values. Correct each standard and unknown reading for the absorbance of the blank. Plot corrected absorbance readings versus standard concentrations, µg Mo/mL, and from the curve determine the unknown concentration based upon its absorbance reading. From this value, determine the total amount of molybdenum present in the sample and the percent molybdenum per gram of dry lubricant.

Molybdenum is sometimes found as a component of *glass* and has been measured in National Bureau of Standards (USA) reference materials by spark source mass spectrometry [13, 14]. Molybdenum has also been measured in laser glass [15].

Methods for determination of molybdenum in various other materials are listed in the tables found with most of the earlier chapters.

References

1. Koirtyohann, S.R.: J. Assoc. Off. Anal. Chem., *55*, 989 (1972)
2. Koirtyohann, S.R., Hamilton, M.: J. Assoc. Off. Anal. Chem., *54*, 787 (1971)
3. Mossoth, E.E.: Adv. Catal., *27*, 265 (1978)
4. Ratnasamy, P., Sivasanker, S.: Catal. Rev., *22*, 401 (1980)
5. Heinemann, H.: Catalysis, Science and Technology, Vol. 1 (Anderson, J.R., Boudart, M. ed.) 1981, Berlin, Springer-Verlag
6. Evans, J.C., Morgan, P.H.: Anal. Chim. Acta, *133*, 329 (1981)
7. Bancroft, G.M., Gupta, R.P., Hardin, A.H., Ternan, M.: Anal. Chem., *51*, 2102 (1979)
8. Labrecque, J.J.: Appl. Spectrosc., *30*, 625 (1976)
9. Lansdown, A.R.: MoS$_2$ Lubrication A Continuation Survey 1975–1976, Nat. Tech. Inform. Ser., Rep. N78-20345/2WK (1978)
10. Tsigdinos, G.A.: Topics Current Chem., *76*, 65 (1978)
11. Kalnin, I.L.: Anal. Chem., *36*, 886 (1964)
12. Julietti, R.J., Wilkinson, J.A.: Analyst (London), *93*, 797 (1968)
13. Bergholz, J., Luck, J., Möller, P., Szachi, W.: Fresenius' Z. Anal. Chem., *269*, 121 (1974)
14. Moore, L.J., Machlan, L.A., Shields, W.R., Garner, E.L.: Anal. Chem., *46*, 1082 (1974)
15. Desyatkova, M.A., Karantseva, A.E.: Nauchn. Tr. Gos. Nauchno-Issled. Proektn. Inst. Redkomet. Prom.-sti, *100*, 43 (1980); Chem. Abstr., *94*, 57524u (1981)

Acknowledgement

Mrs. M. Reger assisted with preparation of references and Mrs. S. Flick and Mrs. S. Koralewski typed the manuscript. The author thanks these people and also his wife, Nancy, for their many kindnesses while writing proceeded.

G.A. Parker

Author Index

Subject Index